Damage Mechanics with Finite Elements

Springer
Berlin
Heidelberg
New York
Barcelona
Hong Kong
London
Milan
Paris
Tokyo

P. I. Kattan · G. Z. Voyiadjis

Damage Mechanics with Finite Elements

Practical Applications with Computer Tools

With 28 Figures

Springer

Assistant Professor P. I. Kattan
Dept. of Civil Engineering
Applied Science University
Amman 11931
Jordan
e-mail: pkattan@nol.com.jo

Boyd Professor G. Z. Voyiadjis
Dept. of Civil Engineering
Louisiana State University
Baton Rouge, LA 70803
U.S.A.
e-mail: voyiadjis@eng.lsu.edu

ISBN 3-540-42279-X Springer-Verlag Berlin Heidelberg New York

Library of Congress Cataloging-in-Publication Data
Kattan, Peter Issa, 1961- Damage mechanics with finite elements : practical applications with computer tools / P.I. Kattan, G.Z. Voyiadjis. p. cm.
Includes bibliographical references and index.
ISBN 354042279X
1. Continuum damage mechanics. 2. Finite element method. I. Voyiadjis, G.Z. II. Title
TA409.K38 2001 620.1'123—dc21 2001042877

Springer-Verlag Berlin Heidelberg New York
a member of BertelsmannSpringer Science+Business Media GmbH

http://www.springer.de

Typesetting: Medio Technologies AG, Berlin
Cover Design: Medio Technologies AG, Berlin

Printed on acid-free paper SPIN 10843701 62/3020 hu - 5 4 3 2 1 0 -

Lovingly Dedicated to My Parents, Issa and Mary

Peter I. Kattan

Lovingly Dedicated to My Wife Christina, whose love, understanding and support made this possible

George Z. Voyiadjis

Preface

The major goal of this book is to present the implementation of some damage models with finite elements. The damage models are based on the principles of continuum damage mechanics and the effective stress concept. Several books have appeared recently on damage mechanics but are mostly theoretical in nature. Alternatively, this book provides a complete finite element program that includes the effects of damage.

The book consists of two parts. Part I includes two chapters mainly reviewing topics from finite element analysis and continuum damage mechanics. The reader is cautioned that the material contained in this part is introductory – other references must be consulted for the theoretical aspects of these topics. For a complete theoretical treatment of the subject, the reader is referred to the book *Advances in Damage Mechanics: Metals and Metal Matrix Composites* by Voyiadjis and Kattan, published in 1999. In Part II the finite element program DNA is introduced in three chapters. DNA stands for "Damage Nonlinear Analysis". The program can be used for the analysis of elasto-plastic material behavior including the effects of damage within the framework of damage mechanics. Two versions of DNA are presented – one for small strain analysis and one for finite strain analysis. The program makes extensive calls to a library of tensor operations developed by the authors. The tensor library is extensively outlined in the last chapter of the book.

The book includes a CD-ROM containing both the executable versions of DNA and Fortran source code. Both the small strain and large strain versions are included. The compiled versions will run on personal computers with Windows 98 or 2000 (including Windows NT). The program is designed to run in DOS mode. However, a special Windows program (or interface) is also available on the CD-ROM that will help run DNA through Windows.

The authors wish to acknowledge the help of Mehrdad Foroozesh, Anthony Venson, and Taehyo Park for their significant contributions to DNA. The authors wish also to acknowledge the help of Nigel K. Clarke, Rashed K. Abu Al-Rub, and Robert J. Dorgan for typing and editing parts of the book. The authors also wish to thank their families for their patience and support throughout the writing of this book. Also, the support of Springer-Verlag in the publication of this book is greatly appreciated.

June 2001

Peter I. Kattan
George Z. Voyiadjis

Contents

1 Damage Mechanics

Kachanov (1958) pioneered the subject of continuum damage mechanics by introducing the concept of effective stress. This concept is based on considering a fictitious undamaged configuration of a body and comparing it with the actual damaged configuration. He originally formulated his theory using simple uniaxial tension. Following Kachanov's work researchers in different fields applied continuum damage mechanics to their areas in fields like brittle materials (Krajcinovic and Foneska, 1981; Krajcinovic, 1988) and ductile materials (Lemaitre, 1984, 1985, 1986; Kachanov, 1986; Murakami, 1988) . In the 1990's applications of continuum damage mechanics to plasticity and composite materials have appeared (Voyiadjis and Kattan, 1990, 1993, 1999; Kattan and Voyiadjis, 1990, 1993a, 1993b, 1996; Voyiadjis and Venson, 1995; Voyiadjis and Thiagarajan, 1996; Voyiadjis and Park, 1997a, 1997b).

In the first part of this chapter (Sections 1.1–1.3) the principles of continuum damage mechanics are applied to the problem of uniaxial tension with damage in the form of both voids and cracks. The effective stress concept is used as originally proposed by Kachanov. Isotropic damage is assumed throughout Section 1.2 of the formulation. A consistent mathematical derivation is presented for the decomposition of the damage tensor into void damage and crack damage components. The derivation is performed using two different methods. It is shown that both methods give exactly the same result.

In Section 1.3 generalization of the derivation is made for the general case of anisotropic damage in three dimensions. It is shown that the components of the tensorial crack and void damage variables are not independent of each other. This seems to prove that a coupling exists between these two types of damage. This may be obvious from the physics of the problem but a rigorous mathematical proof is provided. Furthermore, the exact relations showing the dependence between the components of the crack and void damage variables are derived.

1.1 Review of Continuum Damage Mechanics

The principles of continuum damage mechanics are first reviewed for the case of uniaxial tension. In this case, isotropic damage is assumed throughout. Consider a cylindrical bar subjected to a uniaxial tensile force T as shown in Figure 1.1a. The cross-sectional area of the bar is A and it is assumed that both voids and cracks appear as damage in the bar. The uniaxial stress σ in the bar is found easily from the formula $T = \sigma A$. In order to use the principles of continuum damage mechanics, we consider a fictitious undamaged configuration of the bar as shown in Figure 1.1b. In this configuration all types of damage, including both voids and cracks, are removed from the bar. The effective cross-sectional area of the bar in this configuration is denoted by $\bar{A}$ and the effective uniaxial stress is $\bar{\sigma}$. The bars in both the damaged configuration and the effective undamaged configuration are subjected to the same tensile force T. Therefore, considering the effective undamaged configuration, we have the formula $T = \bar{\sigma}\bar{A}$. Equating the two expression of T obtained from both configurations, one obtains the following expression for the effective uniaxial stress $\bar{\sigma}$:

$$\bar{\sigma} = \frac{A}{\bar{A}}\sigma \tag{1.1}$$

Next, one uses the definition of the damage variable ϕ as originally proposed by Kachanov [1]:

$$\phi = \frac{A - \bar{A}}{A} \tag{1.2}$$

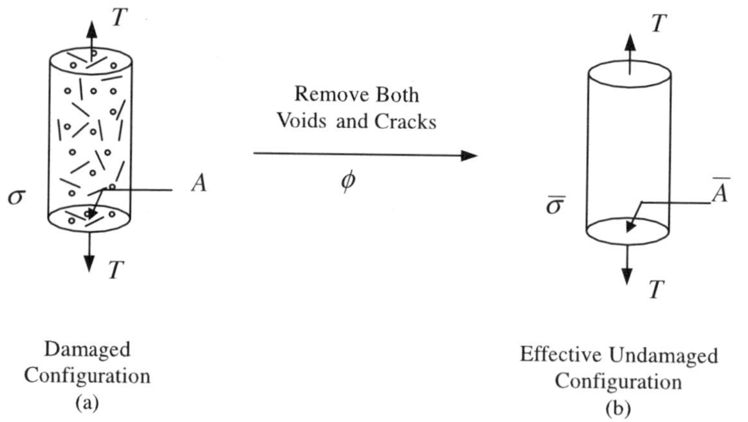

Fig. 1.1. A Cylindrical Bar Subjected to Uniaxial Tension: Both Voids and Cracks Are Removed Simultaneously.

Thus the damage variable is defined as the ratio of the total area of voids and cracks to the total area. Its value ranges from zero (for the case of an undamaged specimen) to one (for the case of complete rupture). Substituting for $A/\overline{A}$ from Equation (1.2) into Equation (1.1), one obtains the following expression for the effective uniaxial stress:

$$\overline{\sigma} = \frac{\sigma}{1 - \phi} \tag{1.3}$$

Equation (1.3) above was originally derived by Kachanov (1958). It is clear from Equation (1.3) that the case of complete rupture ($\phi = 1$) is unattainable because the damage variable ϕ is not allowed to take the value 1 in the denominator.

1.2 Decomposition of the Damage Variable in One Dimension

The principles of continuum damage mechanics are now applied to the problem of decomposition of the damage tensor in a damaged uniaxial bar subjected to a tensile force *T*. Isotropic damage is assumed throughout the formulation. It is also assumed that the damaged state is defined by voids and cracks only. Therefore, the cross-sectional area A of the damaged bar can be decomposed as follows:

$$A = \bar{A} + A^v + A^c \tag{1.4}$$

where A^v is the total area of voids in the cross-section and A^c is the total area of cracks (measured lengthwise) in the cross-section[1]. In addition to the total damage variable ϕ, the two damage variables ϕ^v and ϕ^c are introduced to represent the damage due to voids and cracks, respectively. Our goal is to find a representation for the total damage variable ϕ in terms of ϕ^v and ϕ^c. In order to do this, we need to theoretically separate the damage due to voids and cracks when constructing the effective undamaged configuration. This separation can be performed in two different methods. We can start by removing the voids only, then we remove the cracks separately, or we can start by removing the cracks only, then we can remove the voids separately. The detailed formulation based on each of these two methods is discussed below and is shown schematically in Figures 1.2 and 1.3. It is emphasized that this separation of voids and cracks is theoretical in the sense that it is an acceptable method of mathematical analysis and has no physical basis. In fact, the physics of the problem indicates a coupling between the two damage mechanisms which is apparent in the next Section for the general three-dimensional case.

1 The superscripts "v" and "c" denote voids and cracks, respectively.

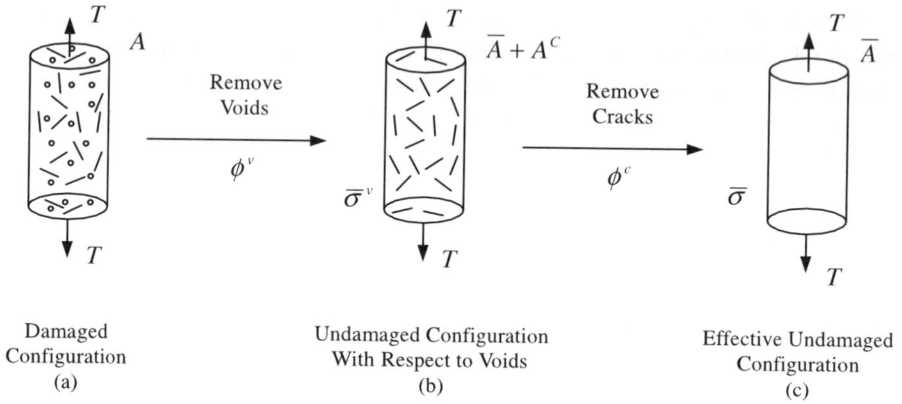

Fig. 1.2. A Cylindrical Bar Subjected to Uniaxial Tension: Voids Are Removed First Then Followed by Cracks.

In the first method, we first remove the voids only from the damaged configuration shown in Figure 1.2a. In this way we obtain the damaged configuration shown in Figure 1.2b which contains damage due to cracks only. This is termed the undamaged configuration with respect to voids. The cross-sectional area of the bar in this configuration is clearly $\bar{A} + A^c$ while the uniaxial stress is denoted by $\bar{\sigma}^v$. The total tensile force T in this configuration is then given by $T = \bar{\sigma}^v(\bar{A} + A^c)$. This expression is equated to the total tensile force $T = \sigma A$ in the damaged configuration from which we obtain:

$$\bar{\sigma}^v = \frac{A}{\bar{A} + A^c}\sigma \tag{1.5}$$

The damage variable ϕ^v due to voids is defined by the ratio A^v/A. Substituting for A^v from Equation (1.4), one obtains:

$$\phi^v = 1 - \frac{\bar{A} + A^c}{A} \tag{1.6}$$

Substituting Equation (1.6) into Equation (1.5), one obtains the following relation between $\bar{\sigma}^v$ and σ:

$$\bar{\sigma}^v = \frac{\sigma}{1 - \phi^v} \tag{1.7}$$

The similarity between Equations (1.7) and (1.3) is very clear. The next step involves removing the cracks from the intermediate configuration in order to obtain the effective undamaged configuration shown in Figure 1.2c. Equating the previous expression for the tensile force $T = \bar{\sigma}^v(\bar{A} + A^c)$ with the tensile force $T = \bar{\sigma}\bar{A}$ in the effective undamaged configuration, one obtains:

$$\bar{\sigma} = \bar{\sigma}^v \left(1 + \frac{A^c}{\bar{A}}\right) \tag{1.8}$$

The damage variable ϕ^c due to cracks is now defined by the ratio:

$$\phi^c = \frac{A^c}{\bar{A} + A^c} \tag{1.9}$$

Substituting Equation (1.9) into Equation (1.8) and simplifying, one obtains the following relation between $\bar{\sigma}$ and $\bar{\sigma}^v$:

$$\bar{\sigma} = \frac{\bar{\sigma}^v}{1 - \phi^c} \tag{1.10}$$

Finally, we substitute Equation (1.7) into Equation (1.10) to obtain the sought relationship between σ and $\bar{\sigma}$:

$$\bar{\sigma} = \frac{\sigma}{(1 - \phi^v)(1 - \phi^c)} \tag{1.11}$$

The above relation represents a formula for the effective stress in terms of the separate damage variables due to voids and cracks.

The same result can be obtained by reversing the order of removal of voids and cracks. In the second method, one first removes the cracks only from the damaged configuration shown in Figure 1.3a. In this way we obtain the damaged configuration shown in Figure 1.3b which contains damage due to voids only. This is termed the undamaged configuration with respect to

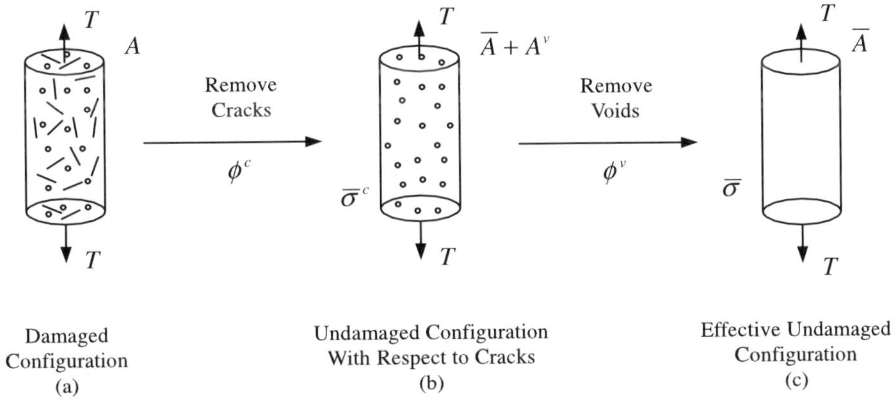

Fig. 1.3. A Cylindrical Bar Subjected to Uniaxial Tension: Cracks Are Removed First Followed by Voids.

cracks. The cross-sectional area of the bar in this configuration is clearly $\bar{A} + A^v$ while the uniaxial stress is denoted by $\bar{\sigma}^c$. The total tensile force T in this configuration is then given by $T = \bar{\sigma}^c(\bar{A} + A^v)$. This expression is equated to the total tensile force $T = \sigma A$ in the damaged configuration from which we obtain:

$$\bar{\sigma}^c = \frac{A}{\bar{A} + A^v}\sigma \tag{1.12}$$

The damage variable ϕ^c due to cracks is defined by the ratio A^c/A. Substituting for A^c from Equation (1.4), one obtains:

$$\phi^c = 1 - \frac{\bar{A} + A^v}{A} \tag{1.13}$$

Substituting Equation (1.13) into Equation (1.12), one obtains the following relation between $\bar{\sigma}^c$ and σ:

$$\bar{\sigma}^c = \frac{\sigma}{1 - \phi^c} \tag{1.14}$$

The similarity between Equations (1.14), (1.7) and (1.3) is very clear. The next step involves removing the voids from the intermediate configuration in order to obtain the effective undamaged configuration shown in Figure 1.3c. Equating the previous expression for the tensile force $T = \bar{\sigma}^c(\bar{A} + A^v)$ with the tensile force $T = \bar{\sigma}\bar{A}$ in the effective undamaged configuration, one obtains:

$$\bar{\sigma} = \bar{\sigma}^c\left(1 + \frac{A^v}{\bar{A}}\right) \tag{1.15}$$

The damage variable ϕ^v due to voids is now defined by the ratio:

$$\phi^v = \frac{A^v}{\bar{A} + A^v} \tag{1.16}$$

Substituting Equation (1.16) into Equation (1.15) and simplifying, one obtains the following relation between $\bar{\sigma}$ and $\bar{\sigma}^c$:

$$\bar{\sigma} = \frac{\bar{\sigma}^c}{1 - \phi^v} \tag{1.17}$$

Finally, we substitute Equation (1.14) into Equation (1.17) to obtain the sought relationship between σ and $\bar{\sigma}$:

$$\bar{\sigma} = \frac{\sigma}{(1 - \phi^c)(1 - \phi^v)} \tag{1.18}$$

It is clear that the above relation between the two stresses in the damaged and the effective configurations is exactly the same relation obtained using the first method, i.e. Equation (1.11). Thus both methods of constructing the effective undamaged configuration give the same relation between the stresses in the respective configurations. In this way, the decomposition of the damage tensor has been completed for the one-dimensional case. In order to derive the final result, we compare either Equation (1.11) or Equation (1.18) with the total damage appearing in Equation (1.3). Equating the denominators on the right-hand-side of these Equations, we can easily obtain the formula:

$$1 - \phi = (1 - \phi^v)(1 - \phi^c) \tag{1.19}$$

Equation (1.19) represents the general form for the decomposition of the damage variable into its two respective components, ϕ^v and ϕ^c. The result can be further simplified by expanding Equation (1.19) and simplifying to obtain:

$$\phi = \phi^v + \phi^c - \phi^v \phi^c \tag{1.20}$$

Equation (1.20) gives a very clear picture of how the total damage variable ϕ can be decomposed into a damage variable ϕ^v due to voids and a damage variable ϕ^c due to cracks. It is also clear that Equation (1.20) satisfies the constraint $0 \leq \phi \leq 1$ whenever each of the other two damage variables satisfies it. It is also clear that when damage in the material is produced by voids only ($\phi^c = 0$), then $\phi = \phi^v$. Alternatively, $\phi = \phi^c$ when damage in the material is produced by cracks only ($\phi^v = 0$).

1.3 General Decomposition of the Damage Variable in Three Dimensions

For the general case of three dimensional deformation and damage, the effective stress tensor $\bar{\sigma}_{ij}$ is given by the following transformation (Voyiadjis and Kattan, 1999):

$$\bar{\sigma}_{ij} = M_{ijkl} \sigma_{kl} \tag{1.21}$$

where the fourth order tensor M_{ijkl} is given by the following 6×6 matrix representation (Voyiadjis and Kattan, 1999):

$$[M]=\frac{1}{2\nabla}\begin{bmatrix} 2\psi_{22}\psi_{33}-2\phi_{23}^2 & 0 & 0 & 2\phi_{13}\phi_{23}+2\phi_{12}\psi_{33} & 2\phi_{12}\phi_{23}+2\phi_{13}\psi_{22} & 0 \\ 0 & 2\psi_{11}\psi_{33}-2\phi_{13}^2 & 0 & 2\phi_{13}\phi_{23}+2\phi_{12}\psi_{33} & 0 & 2\phi_{12}\phi_{13}+2\phi_{23}\psi_{11} \\ 0 & 0 & 2\psi_{11}\psi_{22}-2\phi_{12}^2 & 0 & 2\phi_{12}\phi_{23}+2\phi_{13}\psi_{22} & 2\phi_{12}\phi_{13}+2\phi_{23}\psi_{11} \\ \phi_{13}\phi_{23}+\phi_{12}\psi_{33} & \phi_{13}\phi_{23}+\phi_{12}\psi_{33} & 0 & \psi_{22}\psi_{33}+\psi_{11}\psi_{33}-\phi_{23}^2-\phi_{13}^2 & \phi_{12}\phi_{13}+\phi_{23}\psi_{11} & \phi_{12}\phi_{23}+\phi_{13}\psi_{22} \\ \phi_{12}\phi_{23}+\phi_{13}\psi_{22} & 0 & \phi_{12}\phi_{23}+\phi_{13}\psi_{22} & \phi_{12}\phi_{13}+\phi_{23}\psi_{11} & \psi_{22}\psi_{33}+\psi_{11}\psi_{22}-\phi_{23}^2-\phi_{13}^2 & \phi_{13}\phi_{23}+\phi_{12}\psi_{33} \\ 0 & \phi_{12}\phi_{13}+\phi_{23}\psi_{11} & \phi_{12}\phi_{13}+\phi_{23}\psi_{11} & \phi_{12}\phi_{23}+\phi_{13}\psi_{22} & \phi_{13}\phi_{23}+\phi_{12}\psi_{33} & \psi_{11}\psi_{33}+\psi_{11}\psi_{22}-\phi_{13}^2-\phi_{12}^2 \end{bmatrix} \tag{1.22}$$

where ∇ is given by:

$$\nabla=\psi_{11}\psi_{22}\psi_{33}-\phi_{23}^2\psi_{11}-\phi_{13}^2\psi_{22}-\phi_{12}^2\psi_{33}-2\phi_{12}\phi_{23}\phi_{13} \tag{1.23}$$

and $\psi_{ij}=\delta_{ij}-\phi_{ij}$ (δ_{ij} is the Kronecker delta).

Repeating the same procedure that was employed in the one-dimensional case, we can derive the general decomposition for the case of three dimensions. Removing the voids first, then removing the cracks and applying Equation (1.21), we obtain the following two Equations:

$$\bar{\sigma}_{mn}^{v}=M_{mnkl}^{v}\sigma_{kl} \tag{1.24}$$

$$\bar{\sigma}_{ij}=M_{ijmn}^{c}\bar{\sigma}_{mn}^{v} \tag{1.25}$$

Equations (1.24) and (1.25) correspond to Equations (1.7) and (1.10) of the one-dimensional case. Substituting Equation (1.24) into Equation (1.25), we obtain:

$$\bar{\sigma}_{ij}=M_{ijmn}^{c}M_{mnkl}^{v}\sigma_{kl} \tag{1.26}$$

Comparing Equation (1.26) with Equation (1.21), we obtain the desired general deomposition:

$$M_{ijkl} = M^c_{ijmn} M^v_{mnkl} \tag{1.27}$$

Alternatively, removing the cracks first, then removing the voids and applying Equation (1.21), we obtain the following two Equations:

$$\bar{\sigma}^c_{mn} = M^c_{mnkl} \sigma_{kl} \tag{1.28}$$

$$\bar{\sigma}_{ij} = M^v_{ijmn} \bar{\sigma}^c_{mn} \tag{1.29}$$

Equations (1.28) and (1.29) correspond to Equations (1.14) and (1.17) of the one-dimensional case. Substituting Equation (1.28) into Equation (1.29), we obtain:

$$\bar{\sigma}_{ij} = M^v_{ijmn} M^c_{mnkl} \sigma_{kl} \tag{1.30}$$

Comparing Equation (1.30) with Equation (1.21), we again obtain the desired general decomposition:

$$M_{ijkl} = M^v_{ijmn} M^c_{mnkl} \tag{1.31}$$

It is emphasized that Equations (1.27) and (1.31) are equivalent, i.e. both are valid representations of the decomposition of the damage effect tensor M_{ijkl}. By carrying out the tensorial multiplications explicitly using Equation (1.22), one obtains exactly identical results.

Using the 6×6 matrix representation of Equation (1.22), it is clear that Equations (1.27) and (1.31) reduce to the following form:

$$[M] = [M]^c [M]^v = [M]^v [M]^c \tag{1.32}$$

where both $[M]^v$ and $[M]^c$ can be represented using the 6×6 matrix of Equation (1.22) by replacing ϕ_{ij} with ϕ^c_{ij} and ϕ^v_{ij}, respectively. The same thing holds for ψ_{ij}.

The matrix multiplications in Equation (1.32) are performed using the computer algebra program MAPLE. Performing the first multiplication $[M]^c[M]^v$ and equating the result with Equation (1.22), we obtain 36 Equations between the components of the damage variables ϕ_{ij}, ϕ^v_{ij}, and ϕ^c_{ij}. We also obtain another 36 Equations (for a total of 72 Equations) when we perform the second multiplication $[M]^v$ $[M]^c$. It is noted that many of these Equations are identical or can be shown to be identical. We have been able to reduce this huge set of Equations to 18 independent Equations. Furthermore, these 18 Equations have been classified into two categories. The first category includes the following nine Equations which represent the general decomposition of the damage tensor components ϕ_{ij} into the damage tensor components ϕ^c_{ij} due to cracks and the damage tensor components ϕ^v_{ij} due to voids.

$$\frac{1}{\nabla}(\psi_1 1\psi_{22} - \phi^2_{12}) = \frac{1}{\nabla^c\nabla^v}(\psi^c_{11}\psi^c_{22} - \phi^{c^2}_{12})(\psi^v_{11}\psi^v_{22} - \phi^{v^2}_{12}) \tag{1.33a}$$

$$\frac{1}{\nabla}(\psi_{11}\psi_{33} - \phi^2_{13}) = \frac{1}{\nabla^c\nabla^v}(\psi^c_{11}\psi^c_{33} - \phi^{c^2}_{13})(\psi^v_{11}\psi^v_{33} - \phi^{v^2}_{13}) \tag{1.33b}$$

$$\frac{1}{\nabla}(\psi_{22}\psi_{33} - \phi^2_{23}) = \frac{1}{\nabla^c\nabla^v}(\psi^c_{22}\psi^c_{33} - \phi^{c^2}_{23})(\psi^v_{22}\psi^v_{33} - \phi^{v^2}_{23}) \tag{1.33c}$$

$$\begin{aligned}\frac{1}{\nabla}(\phi_{13}\phi_{23} + \phi_{12}\psi_{33}) &= \frac{1}{\nabla^c\nabla^v}[\left(\psi^c_{22}\psi^c_{33} - \phi^{c^2}_{23}\right)(\phi^v_{13}\phi^v_{23} + \phi^v_{12}\psi^v_{33}) \\ &+ (\phi^c_{13}\phi^c_{23} + \phi^c_{12}\psi^c_{33})\left(\psi^v_{22}\psi^v_{33} + \psi^v_{11}\psi^v_{33} - \phi^{v^2}_{23} - \phi^{v^2}_{13}\right)]\end{aligned} \tag{1.33d}$$

$$\begin{aligned}\frac{1}{\nabla}(\phi_{12}\phi_{23} + \phi_{13}\psi_{22}) &= \frac{1}{\nabla^c\nabla^v}[\left(\psi^c_{22}\psi^c_{33} - \phi^{c^2}_{23}\right)(\phi^v_{12}\phi^v_{23} + \phi^v_{13}\psi^v_{22}) \\ &+ (\phi^c_{12}\phi^c_{23} + \phi^c_{13}\psi^c_{22})\left(\psi^v_{22}\psi^v_{33} + \psi^v_{11}\psi^v_{33} - \phi^{v^2}_{23} - \phi^{v^2}_{12}\right)]\end{aligned} \tag{1.33e}$$

$$\begin{aligned}\frac{1}{\nabla}(\phi_{12}\phi_{13} + \phi_{23}\psi_{11}) &= \frac{1}{\nabla^c\nabla^v}[\left(\psi^c_{11}\psi^c_{33} - \phi^{c^2}_{13}\right)(\phi^v_{12}\phi^v_{13} + \phi^v_{23}\psi^v_{11}) \\ &+ (\phi^c_{12}\phi^c_{13} + \phi^c_{23}\psi^c_{11})\left(\psi^v_{11}\psi^v_{33} + \psi^v_{11}\psi^v_{22} - \phi^{v^2}_{13} - \phi^{v^2}_{12}\right)]\end{aligned} \tag{1.33f}$$

$$\begin{aligned}&\frac{1}{\nabla}(\psi_{11}\psi_{22} + \psi_{22}\psi_{33} - \phi^2_{12} - \phi^2_{23}) \\ &= \frac{1}{\nabla^c\nabla^v}(\psi^c_{11}\psi^c_{22} + \psi^c_{22}\psi^c_{33} - \phi^{c^2}_{12} - \phi^{c^2}_{23})(\psi^v_{11}\psi^v_{22} + \psi^v_{22}\psi^v_{33} - \phi^{v^2}_{12} - \phi^{v^2}_{23})\end{aligned} \tag{1.33g}$$

$$\begin{aligned}&\frac{1}{\nabla}(\psi_{11}\psi_{33} + \psi_{22}\psi_{33} - \phi^2_{13} - \phi^2_{23}) \\ &= \frac{1}{\nabla^c\nabla^v}(\psi^c_{11}\psi^c_{33} + \psi^c_{22}\psi^c_{33} - \phi^{c^2}_{13} - \phi^{c^2}_{23})(\psi^v_{11}\psi^v_{33} + \psi^v_{22}\psi^v_{33} - \phi^{v^2}_{13} - \phi^{v^2}_{23})\end{aligned} \tag{1.33h}$$

$$\begin{aligned}&\frac{1}{\nabla}(\psi_{11}\psi_{22} + \psi_{11}\psi_{33} - \phi^2_{12} - \phi^2_{23}) \\ &= \frac{1}{\nabla^c\nabla^v}(\psi^c_{11}\psi^c_{22} + \psi^c_{11}\psi^c_{33} - \phi^{c^2}_{12} - \phi^{c^2}_{13})(\psi^v_{11}\psi^v_{22} + \psi^v_{11}\psi^v_{33} - \phi^{v^2}_{12} - \phi^{v^2}_{23})\end{aligned} \tag{1.33i}$$

where $\psi^c_{ij} = \delta_{ij} - \phi^c_{ij}$ and $\psi^v_{ij} = \delta_{ij} - \phi^v_{ij}$.

It is clear from the above set of Equations that the decomposition is not explicit. A set of simultaneous Equations needs to be solved for the general three-dimensional decomposition. The second category of Equations includes the remaining nine Equations as follows:

$$(\phi_{13}^c\phi_{23}^c + \phi_{12}^c\psi_{33}^c)(\phi_{13}^v\phi_{23}^v + \phi_{12}^v\psi_{33}^v) = 0 \tag{1.34a}$$

$$(\phi_{12}^c\phi_{23}^c + \phi_{13}^c\psi_{22}^c)(\phi_{12}^v\phi_{23}^v + \phi_{13}^v\psi_{22}^v) = 0 \tag{1.34b}$$

$$(\phi_{12}^c\phi_{13}^c + \phi_{23}^c\psi_{11}^c)(\phi_{12}^v\phi_{13}^v + \phi_{23}^v\psi_{11}^v) = 0 \tag{1.34c}$$

$$(\phi_{12}^c\phi_{13}^c + \phi_{23}^c\psi_{11}^c)(\phi_{13}^v\phi_{23}^v + \phi_{12}^v\psi_{33}^v) = 0 \tag{1.34d}$$

$$(\phi_{12}^c\phi_{23}^c + \phi_{13}^c\psi_{22}^c)(\phi_{13}^v\phi_{23}^v + \phi_{12}^v\psi_{33}^v) = 0 \tag{1.34e}$$

$$(\phi_{13}^c\phi_{23}^c + \phi_{12}^c\psi_{33}^c)(\phi_{12}^v\phi_{23}^v + \phi_{13}^v\psi_{22}^v) = 0 \tag{1.34f}$$

$$(\phi_{12}^c\phi_{23}^c + \phi_{13}^c\psi_{22}^c)(\phi_{12}^v\phi_{13}^v + \phi_{23}^v\psi_{11}^v) = 0 \tag{1.34g}$$

$$(\phi_{12}^c\phi_{13}^c + \phi_{23}^c\psi_{11}^c)(\phi_{12}^v\phi_{23}^v + \phi_{13}^v\psi_{22}^v) = 0 \tag{1.34h}$$

$$(\phi_{13}^c\phi_{23}^c + \phi_{12}^c\psi_{33}^c)(\phi_{12}^v\phi_{13}^v + \phi_{23}^v\psi_{11}^v) = 0 \tag{1.34i}$$

It is clear that the above Equations do not contain any components of the total damage tensor ϕ_{ij}. This set of Equations relate only the components of the two damage tensors ϕ_{ij}^c and ϕ_{ij}^v. It is concluded that the above set of nine Equations represent the exact coupling between the two damage mechanisms of voids and cracks. Although this coupling may be obvious based on the physics of the problem, a rigorous mathematical proof has been given for it. Furthermore, the coupling Equations (1.34) have not appeared before in the literature.

Finally it is noted that for the special case of one-dimensional damage, Equations (1.33) reduce to the simple decomposition shown in Equations (1.19) and (1.20) while the coupling Equations (1.34) reduce to zero.

1.4 Damage of Fiber-Reinforced Composite Materials with Micromechanical Characterization

Fiber-reinforced composite materials play an important role in the industry today through the design and manufacture of advanced materials capable of attaining higher stiffness/density and strength/density ratios. Of particular importance is the problem of damage initiation and evolution in fiber-reinforced metal matrix composite plates. Although the literature is rich in new developments in the composite materials technology, it lacks tremendously a consistent analysis of damage mechanisms in composite materials.

In the analysis of composite materials, one can follow a continuum approach or a micromechanical approach. In the continuum approach, the composite material is treated as an orthotropic or transversely isotropic medium. Then the classical Equations of orthotropic elasticity are used in the analysis (Talreja, 1985, 1986; Christensen, 1988, 1990). No distinction is made between the matrix and fibers in this approach and therefore, this approach lacks the capability to account for local effects and especially the effects of the matrix-fiber interaction. There were some attempts to include damage using the continuum approach (Talreja, 1985; Shen *et al.*, 1985; Lene, 1986). However, these attempts lack the distinction between matrix and fiber damage or damage due to the matrix-fiber interaction.

During the past two decades, researchers have been using micromechanical methods in the analysis of composite materials. The advantages of using such methods are that local effects can be accounted for and different damage mechanisms can be identified. Hill (1965, 1972) introduced volume averages of stress and strain increments in the matrix and fibers and introduced certain concentration factors to relate these volume averages to the overall uniform increments. Dvorak and Bahei-El-Din (1979, 1982, 1987) used Hill's method to analyze the elasto-plastic behavior of fiber-reinforced composite materials. They considered elastic fibers embedded in an elasto-plastic matrix and identified two distinct deformation modes; matrix dominated and fiber-dominated. They concluded that the fiber-dominated mode is general in the sense that it can be treated as a general case of plastic deformation of a heterogeneous medium.

A thermomechanical constitutive theory has been recently proposed by Allen and Harris (1987a, 1987b) to analyze distributed damage in elastic composites. In particular, the problem of matrix cracking has been extensively studied in the literature (Dvorak, *et al.*, 1985; Dvorak and Laws, 1987; Laws and Dvorak, 1987; Allen, *et al.*, 1988; Lee, *et al.*, 1989).

The theoretical formulation presented here is based on the concept of effective stress that was originally proposed by Kachanov (1958). The pioneering work of Kachanov (1958) started what is now known as continuum damage mechanics. Different researchers (Lamaitre, 1985, 1986; Chaboche, 1988; Krajcinovic, 1983, 1984) used continuum damage mechanics to analyze different types of damage in materials ranging from brittle fracture to ductile failure. However, no attempt has been made to use the concepts of continuum damage mechanics to analyze damage in composite materials using the micromechanical approach. It should be mentioned that some researchers (Talreja, 1985) have used it to analyze damage in composite materials using the continuum approach by modeling the composite as a transversely isotropic medium.

In the following Sections (1.4.1–1.4.6), continuum damage mechanics is used with a micromechanical composite model to analyze damage in composite materials. Both overall and local damage variables are introduced to model the overall and local damage effects. Stress and strain concentration factors are derived for the damaged composite. The model is applied in detail to a unidirectional thin lamina that is subjected to uniaxial tension. It is also applied to a unidirectional thin lamina under a state of plane stress. The research presented in these Sections is the three-dimensional generalization of the uniaxial tension model derived previously by the authors (Kattan and Voyiadjis, 1993).

1.4.1 Definitions and Assumptions

Consider a body of fiber-reinforced composite material in the initial undeformed and undamaged configuration C_0. Let C be the configuration of the body that is both damaged and deformed after a set of external agencies act on it. Following the concept of effective stress (Kachanov, 1958; Murakami, 1988), consider a fictitious configuration of the body $\overline{C}$ obtained from C by removing all the damage that the body (both matrix and fibers) has undergone, i.e. $\overline{C}$ is the state of the body after it had only deformed without damage. Assume that the representative volume element in C_0 is statistically homogeneous, and is free of voids and cracks initially. Assume also that the composite is loaded by an overall stress or strain field which is followed by increments of loading. The overall stress or strain fields are assumed to be uniform. The effective overall stress is defined in the configuration $\overline{C}$ as the stress in a perfectly-bonded two-phase composite free of cracks or voids.

The composite material is assumed to consist of elastic fibers and an elastic matrix. The fibers are continuous, aligned, and equally spaced. It is also assumed that the elastic strains are small (infinitesimal). Therefore, the elastic strain tensor can be taken to be the usual engineering elastic strain tensor ε. It is also assumed that there exists an elastic strain energy function such that a linear relation can be used between the Cauchy stress tensor σ and the engineering elastic strain tensor ε. In fact, the tensor rate $\dot{\varepsilon}$ for small elastic deformations is equal to the elastic part of the spatial strain rate tensor $\mathbf{d}$ where second order terms are neglected.

In the following, quantities are defined in the configuration C of the overall composite system. Barred quantities are defined in the configuration $\overline{C}$ of the overall composite system. Only Cartesian tensors are considered in this work with their tensor components denoted by subscripts with the usual summation convention. Quantities with a superscript M or F refer to matrix or fiber related quantities, respectively. The superscript R is used to indicate the matrix or fibers where no distinction between them is necessary. No summation is assumed between a superscript and the corresponding identical subscript. It follows directly that barred quantities with a superscript M or F (or R in general) refer to matrix or fiber related quantities, respectively, in the configuration $\overline{C}$. For example, σ is the composite (overall) Cauchy stress in C, $\bar{\sigma}$ is the effective composite Cauchy stress in $\overline{C}$, σ^M and σ^F are the matrix and fiber stresses in C, respectively, and $\bar{\sigma}^M$ and $\bar{\sigma}^F$ are the effective matrix and fiber stresses in $\overline{C}$, respectively.

The constitutive model is first formulated in the configuration $\overline{C}$ of the composite system. Then the hypothesis of elastic energy equivalence (Sidoroff, 1981) is used to transform the model into the configuration C of the composite system. In this hypothesis, it is assumed that the elastic energy for a damaged material is equivalent in form to that of the undamaged material except that the stress is replaced by the effective stress in the energy formulation. For this purpose, certain transformation Equations are derived for the composite (overall) stresses and strains between the configurations C and $\overline{C}$.

In the formulation, the Eulerian reference system is used, i.e., all quantities are based on spatial coordinates. In Section 1.4.2, the necessary Equations relating local and overall quantities of the composite system are presented. The continuum damage mechanics Equations are then derived in Section 1.4.3. Then the constitutive Equations are derived in Section 1.4.4.

1.4.2 Composite Analysis

In this Section, the relations between the local (matrix and fiber) and overall (composite) relations are presented in the configuration $\overline{C}$. The analysis is based on the model given by Dvorak and Bahei-El-Din (1982, 1987) and Bahei-El-Din and Dvorak (1989) utilizing a representative volume element that is statistically homogeneous with uniform overall fields of stress or strain. In this case, the composite system consists of an elastic metal matrix reinforced by elastic, continuous aligned fibers.

In the configuration $\overline{C}$, the effective stress tensor $\bar{\sigma}^R$ is related to the effective composite stress tensor $\bar{\sigma}$ by

$$\bar{\sigma}^R_{ij} = B^R_{ijkl}\bar{\sigma}_{kl}, \tag{1.35}$$

where B^R_{ijkl} is a fourth-rank tensor indicating the elastic phase stress concentration factor and the superscript R stands for either M or F. The tensor $\mathbf{B}^R(\mathbf{x})$ depends only on the spatial coordinates $\mathbf{x}$ for the case of elastic deformation. In order to determine $\mathbf{B}^R$, certain assumptions are employed, like the Voigt assumption where the matrix and fibers are assumed to deform equally or the VFD (Vanishing Fiber Diameter) assumption where the fibers are assumed to have vanishing diameters while occupying a finite volume fraction. These two assumptions are discussed briefly among others at the end of Section

1.4.4. The reader should note that the tensor $\mathbf{B}^{\mathrm{R}}$ does not include any damage effect. This is the reason why effective stresses are used in Equation (1.35) rather than the actual stresses. In the sequel, a damage phase stress concentration factor will be derived in terms of $\mathbf{B}^{\mathrm{R}}$ and the damage variables.

As a result of volume integration and averaging of the local stress fields, the following relation is obtained between the local (matrix and fiber) stresses and the overall stress in $\overline{C}$:

$$\bar{\sigma}_{ij} = c^{\mathrm{M}}\bar{\sigma}_{ij}^{\mathrm{M}} + c^{\mathrm{F}}\bar{\sigma}_{ij}^{\mathrm{F}}, \tag{1.36}$$

where c^{M} and c^{F} are the matrix and fiber volume fractions, respectively, given by:

$$c^{\mathrm{R}} = V^{\mathrm{R}}/V, \quad V = V^{\mathrm{M}} + V^{\mathrm{F}}. \tag{1.37}$$

In Equation (1.37), V^{M} and V^{F} are the matrix and fiber volumes, respectively, and V is the total volume of the representative composite element. Using the assumption in Equation (1.36) and substituting the relevant expressions for $\bar{\sigma}^{\mathrm{M}}$ and $\bar{\sigma}^{\mathrm{F}}$ from Equation (1.35), one derives the following relation between the elastic stress concentration factors for the matrix and fibers:

$$c^{\mathrm{M}}B_{ijkl}^{\mathrm{M}} + c^{\mathrm{F}}B_{ijkl}^{\mathrm{F}} = \frac{1}{2}\left(\delta_{ik}\delta_{jl} + \delta_{il}\delta_{jk}\right), \tag{1.38}$$

where δ_{ij} is the Kronecker delta.

It is now seen that once the elastic matrix stress concentration factor $\mathbf{B}^{\mathrm{M}}$ is determined, one can use Equation (1.38) to find the corresponding fiber stress concentration factor $\mathbf{B}^{\mathrm{F}}$. It also follows from the symmetry of the stress tensor and Equation (1.35) that the stress concentration factors $\mathbf{B}^{\mathrm{M}}$ and $\mathbf{B}^{\mathrm{F}}$ are symmetric in the sense $B_{ijkl}^{\mathrm{R}} = B_{jikl}^{\mathrm{R}}$. Although the tensors $\mathbf{B}^{\mathrm{M}}$ and $\mathbf{B}^{\mathrm{F}}$ have other symmetries, this is the only one needed in the derivations that follow.

Next the effective matrix and fiber deviatoric stresses $\bar{\tau}^{\mathrm{M}}$ and $\bar{\tau}^{\mathrm{F}}$, respectively, are directly derived from Equation (1.35) as follows:

$$\bar{\tau}_{ij}^{\mathrm{R}} = F_{ijkl}^{\mathrm{R}}\bar{\sigma}_{kl}, \tag{1.39}$$

where the fourth rank tensor $\mathbf{F}^{\mathrm{R}}$ is given by:

$$F_{ijkl}^{\mathrm{R}} = B_{ijkl}^{\mathrm{R}} - \frac{1}{3}B_{ppkl}^{\mathrm{R}}\delta_{ij}. \tag{1.40}$$

Using the relation (1.40) above, one can derive the following two useful identities:

$$F_{rrkl}^{\mathrm{R}} = 0, \tag{1.41}$$

$$F_{ijkl}^{\mathrm{R}}F_{ijmn}^{\mathrm{R}} = B_{ijkl}^{\mathrm{R}}B_{ijmn}^{\mathrm{R}}. \tag{1.42}$$

In the next part of this Section, the local-overall relationships for the effective strain tensor $\bar{\varepsilon}$ in the configuration $\overline{C}$ are presented. Upon volume integrating and averaging the local stress fields (Dvorak and Bahei-El-Din, 1982, 1987), the following local-overall relation is obtained for the effective spatial strain tensor

$$\bar{\varepsilon}_{ij} = c^{\mathrm{M}} \bar{\varepsilon}_{ij}^{\mathrm{M}} + c^{\mathrm{F}} \bar{\varepsilon}_{ij}^{\mathrm{F}}, \tag{1.43}$$

where the appropriate relations for the effective matrix and fiber strain tensors are used as follows:

$$\bar{\varepsilon}_{ij}^{\mathrm{R}} = A_{ijkl}^{\mathrm{R}} \bar{\varepsilon}_{kl}, \tag{1.44}$$

where A_{ijkl}^{R} is a fourth-rank tensor denoting the elastic phase strain concentration factor. The same remarks outlined earlier about the tensor $\mathbf{B}^{\mathrm{R}}$ apply again to the tensor $\mathbf{A}^{\mathrm{R}}$.

Substituting Equation (1.44) into Equation (1.43), one derives the following relation between the elastic matrix and fiber strain concentration factors:

$$c^{\mathrm{M}} A_{ijkl}^{\mathrm{M}} + c^{\mathrm{F}} A_{ijkl}^{\mathrm{F}} = \frac{1}{2}\left(\delta_{ik}\delta_{jl} + \delta_{il}\delta_{jk}\right), \tag{1.45}$$

It is now clear that once one of the elastic strain concentration factors is determined, Equation (1.45) can be used to determine the other one.

On the other hand, one may start with the quantity $\bar{\sigma}_{ij}\,\bar{\varepsilon}_{ij}$ and expand it using Equations (1.36) and (1.43). Then one substitutes for the local stresses and strains from Equations (1.35) and (1.44) and simplifies to obtain:

$$\left(c^{\mathrm{M}} B_{ijkl}^{\mathrm{M}} + c^{\mathrm{F}} B_{ijkl}^{\mathrm{F}}\right)\left(c^{\mathrm{M}} A_{ijmn}^{\mathrm{M}} + c^{\mathrm{F}} A_{ijmn}^{\mathrm{F}}\right) = \frac{1}{2}(\delta_{km}\delta_{ln}+\delta_{kn}\delta_{ml}). \tag{1.46}$$

The above Equation represents the relation between the stress and strain concentration factors for the matrix and fibers. In view of the relations (1.39) and (1.46), it is clear that Equation (1.46) reduces to an identity. It should be mentioned that once the stress concentration factors B_{ijkl}^{M} and B_{ijkl}^{F} are determined, one can use Equation (1.46) to find a constraint relation between the strain concentration factors A_{ijkl}^{M} and A_{ijkl}^{F}.

1.4.3 Damage Analysis

There are two steps that can be followed in order to develop a continuum damage model for a composite system consisting of fibers and a matrix. First, one considers damage in the overall composite system as a whole continuum. At this step, the model will reflect various types of damage mechanisms such as void growth and coalescence in the matrix, fiber fracture, debonding and delamination, etc. It should be noted that at this step, no distinction is made between these types of damage as they are all reflected through the fourth-rank overall damage effect tensor M_{ijkl}. In the second step, one considers the damage that the matrix and fibers undergo separately such as nucleation and growth of voids and void coalescence for the matrix and fracture for the fibers. In this case, two fourth-rank matrix and fiber damage effect tensors M^{M}_{ijkl} and M^{F}_{ijkl} are introduced that reflect all types of damage that the matrix and fibers undergo. Subsequently, the local-overall relations are used to transform these local damage effects to the whole composite system. Therefore, it is clear that the second step does not account explicitly for such damage mechanisms as debonding or delamination. It is also clear that each step has certain advantages and disadvantages. While the first step accounts for all types of damage in the composite system, it cannot distinguish between them. In contrast the second step provides separate damage analysis of the matrix and fiber material but lacks the ability to account for fiber-matrix interaction damage. Therefore, the aim of the proposed model will be to combine the two aforementioned steps in such a way so as to isolate the various local types of damage.

Following the first step outlined above and utilizing an overall damage effect tensor $\mathbf{M}$ for the whole composite system, the overall effective Cauchy stress tensor $\bar{\sigma}$ is given by

$$\bar{\sigma}_{ij} = M_{ijkl}\,\sigma_{kl}. \tag{1.47}$$

The above relation was first proposed for the uniaxial case by Kachanov (1958) and later generalized to three dimensions by Murakami (1988) and Sidoroff (1981) in the framework of the concept of effective stress. It then follows from Equation (1.47) that the overall effective deviatoric Cauchy stress tensor is given by (Kattan and Voyiadjis, 1990; Voyiadjis and Kattan, 1992)

$$\bar{\tau}_{ij} = N_{ijkl}\,\sigma_{kl}, \tag{1.48}$$

where the fourth-rank tensor $\mathbf{N}$ is given in terms of $\mathbf{M}$ as follows:

$$N_{ijkl} = M_{ijkl} - \frac{1}{3} M_{rrkl}\delta_{ij}. \tag{1.49}$$

Certain useful identities follow directly from Equation (1.49). The main two identities used here are listed below:

$$N_{rrkl} = 0, \tag{1.50}$$

$$N_{ijkl}N_{ijmn} = M_{ijkl}M_{ijmn}. \tag{1.51}$$

Next, the relation between the effective phase stress tensor $\bar{\sigma}^{\mathrm{R}}$ and the overall stress tensor σ is derived. This is done by substituting Equation (1.47) into Equation (1.35). Therefore, one obtains:

$$\bar{\sigma}^{\mathrm{R}}_{ij} = K^{\mathrm{R}}_{ijkl}\,\sigma_{kl}, \tag{1.52}$$

where the fourth-rank matrix tensor K^{R}_{ijkl} is given by:

$$K^{\mathrm{R}}_{ijkl} = B^{\mathrm{R}}_{ijmn} M_{mnkl}. \tag{1.53}$$

From the symmetry of B^{R}_{ijmn} discussed earlier, it follows from Equation (1.53) that the tensor $\mathbf{K}^{\mathrm{R}}$ is symmetric in the sense $K^{\mathrm{R}}_{ijkl} = K^{\mathrm{R}}_{jikl}$. It should be noted that the tensor $\mathbf{K}^{\mathrm{R}}$ has other symmetries, but are not needed in the derivations.

Substituting Equation (1.47) into Equation (1.39), one obtains the following expressions for the effective matrix and fiber deviatoric stress tensors:

$$\bar{\tau}^{\mathrm{R}}_{ij} = R^{\mathrm{R}}_{ijkl}\,\sigma_{kl}, \tag{1.54}$$

where the fourth-rank tensor $\mathbf{R}^{\mathrm{R}}$ is given by:

$$R^{\mathrm{R}}_{ijkl} = F^{\mathrm{R}}_{ijmn} M_{mnkl}. \tag{1.55}$$

Upon examining Equations (1.39), (1.40), (1.54) and (1.55), one concludes that the tensors $\mathbf{F}^{\mathrm{R}}$ and $\mathbf{R}^{\mathrm{R}}$ are symmetric in the sense $F^{\mathrm{R}}_{ijkl} = F^{\mathrm{R}}_{jikl}$ and $R^{\mathrm{R}}_{ijkl} = R^{\mathrm{R}}_{jikl}$. Furthermore, by substituting Equation (1.40) into Equation (1.55) and using Equation (1.53), one can derive the following relation between the tensors $\mathbf{K}^{\mathrm{R}}$ and $\mathbf{R}^{\mathrm{R}}$:

$$R^{\mathrm{R}}_{ijkl} = K^{\mathrm{R}}_{ijkl} - \frac{1}{3} K^{\mathrm{R}}_{rrkl}\delta_{ij}. \tag{1.56}$$

The tensor $\mathbf{R}^{\mathrm{R}}$ satisfies the two identities discussed earlier, namely $R^{\mathrm{R}}_{rrkl} = 0$ and $R^{\mathrm{R}}_{ijkl} R^{\mathrm{R}}_{ijmn} = K^{\mathrm{R}}_{ijkl} K^{\mathrm{R}}_{ijmn}$. All these tensors have other symmetries but are not needed in the derivations.

The overall damage relations for the composite system have now been presented in Equations (1.47) through (1.56). The overall damage effect tensor $\mathbf{M}$ has been introduced to represent all types of damage that the system undergoes. Following the second step discussed at the beginning of this Section, one introduces a phase (local) damage effect tensor $\mathbf{M}^{\mathrm{R}}$ that represents the damage mechanisms in the phase material like nucleation, growth and coalescence of voids for the matrix, and fracture of fibers. Therefore, the following local transformation Equation is assumed to hold for the phase stress tensor

$$\bar{\sigma}^{\mathrm{R}}_{ij} = M^{\mathrm{R}}_{ijkl}\sigma^{\mathrm{R}}_{kl}. \tag{1.57}$$

It now follows directly from Equation (1.57) that

$$\bar{\tau}^{\mathrm{R}}_{ij} = N^{\mathrm{R}}_{ijkl}\sigma^{\mathrm{R}}_{kl}, \tag{1.58}$$

where the fourth-rank tensor $\mathbf{N}^{\mathrm{R}}$ satisfies the relations in Equations (1.49), (1.50) and (1.51). Comparing Equations (1.52) and (1.57) and simplifying, one derives the following relation between the phase stress tensor and the overall stress tensor:

$$\sigma_{ij}^{\mathrm{R}} = \bar{B}_{ijkl}^{\mathrm{R}} \sigma_{kl}, \tag{1.59}$$

where

$$\bar{B}_{ijkl}^{\mathrm{R}} = \left(M_{mnij}^{\mathrm{R}}\right)^{-1} B_{mnpq}^{\mathrm{R}} M_{pqkl}, \tag{1.60}$$

where the inverse W_{ijkl}^{-1} of a fourth-rank tensor W_{ijkl} is defined by $W_{ijmn} W_{ijkl}^{-1} = \delta_{ik}\delta_{jl}$. The fourth-rank tensor $\mathbf{B}^{\mathrm{R}}$ is the damaged phase stress concentration factor that includes geometrical and damage related effects as can be seen from Equation (1.60).

It is now possible to derive the required relationship between the local damage effect tensors $\mathbf{M}^{\mathrm{M}}$ and $\mathbf{M}^{\mathrm{F}}$ and the overall damage effect tensor $\mathbf{M}$. Substituting Equations (1.57) and (1.59) into Equation (1.36) and simplifying, one obtains the desired relation:

$$M_{ijkl} = c^{\mathrm{M}} M_{ijmn}^{\mathrm{M}} \bar{B}_{mnkl}^{\mathrm{M}} + c^{\mathrm{F}} M_{ijmn}^{\mathrm{F}} \bar{B}_{mnkl}^{\mathrm{F}}. \tag{1.61}$$

It is clear that Equation (1.61) relates the local damage experienced by the matrix and fibers to the overall damage of the composite system. The damaged matrix and fiber concentration factors appear in the Equation as well as the matrix and fiber volume fractions. Substituting for $\bar{\mathbf{B}}^{\mathrm{M}}$ and $\bar{\mathbf{B}}^{\mathrm{F}}$ from Equation (1.60) into Equation (1.61), one obtains:

$$M_{ijkl} = c^{\mathrm{M}} B_{ijmn}^{\mathrm{M}} M_{mnkl} + c^{\mathrm{F}} B_{ijmn}^{\mathrm{F}} M_{mnkl}. \tag{1.62}$$

The above Equation is an explicit relation between the effective local concentration factors and the overall damage effect tensor. Examining Equations (1.61) and (1.62) carefully, one concludes that once the local (matrix and fiber) damage mechanisms have been described through the tensors $\mathbf{M}^{\mathrm{M}}$ and $\mathbf{M}^{\mathrm{F}}$, then the overall damage in the composite system can be described which includes the matrix and fiber related damage as well as the damage resulting from the interaction of the two phases such as debonding.

1.4.4 Constitutive Equations

In this Section, the elastic constitutive relation for the damaged composite system will be developed. First, one starts with the overall system and assumes the material obeys generalized Hooke's law in the undamaged configuration $\bar{C}$:

$$\bar{\sigma}_{ij} = E_{ijkl} \bar{\varepsilon}_{kl}, \tag{1.63}$$

where E_{ijkl} is the constant fourth-rank elasticity tensor. The corresponding effective elastic strain energy $\bar{U}$ in this configuration is given by

$$\bar{U} = \frac{1}{2} E_{ijkl} \bar{\varepsilon}_{ij} \bar{\varepsilon}_{kl}. \tag{1.64}$$

One now uses the Legendre transform and applies it to Equation (1.64) in order to derive the following expression for the effective elastic complementary strain energy $\bar{V}$:

$$\bar{V} = \frac{1}{2} E^{-1}_{ijkl} \bar{\sigma}_{ij} \bar{\sigma}_{kl}. \tag{1.65}$$

In the damaged composite configuration C, the elastic constitutive relation takes the form

$$\sigma_{ij} = \bar{E}_{ijkl} \varepsilon_{kl}, \tag{1.66}$$

where the fourth-rank tensor $\bar{E}_{ijkl}$ is no longer constant but depends on the damage effect tensor M_{ijkl}. Using the hypothesis of elastic energy equivalence, by equating the energy in Equations (1.65) in both the damaged and undamaged configurations, i.e. $\bar{V} = V$, one derives the following expression for $\bar{E}_{ijkl}$ (Kattan and Voyiadjis, 1990).

$$\bar{E}_{ijkl} = M^{-1}_{pqkl} E_{rspq} M^{-1}_{rsij}. \tag{1.67}$$

Differentiating Equation (1.65) with respect to $\bar{\sigma}$, and using Equations (1.47), (1.66) and (1.67), one can obtain the following transformation Equation for the overall strain tensor in the configuration $\bar{C}$:

$$\bar{\varepsilon}_{ij} = M^{-1}_{ijpq} \varepsilon_{\mathrm{pq}}. \tag{1.68}$$

The above Equations (1.63) – (1.68) are a brief review of the elastic constitutive relations for a damaged one-phase material and can be used as the overall relations for the composite system. Next, one considers the local stresses and strains in an attempt to formulate local-overall Equations for the damaged composite system.

Similar relations can be shown to exist on the local level, that is, the strain transformation Equations for the matrix and fibers are similar to Equation (1.68) and take the following form:

$$\bar{\varepsilon}^{\mathrm{R}}_{ij} = \left(M^{\mathrm{R}}_{ijmn} \right)^{-1} \varepsilon^{\mathrm{R}}_{mn}. \tag{1.69}$$

One now can derive expressions for the damaged strain concentration factors. Substituting the expressions of the transformation of the overall and local strains of Equations (1.68) and (1.69) into Equation (1.44) and simplifying, one obtains

$$\varepsilon^{\mathrm{R}}_{ij} = \bar{A}^{\mathrm{R}}_{ijkl} \varepsilon_{kl}, \tag{1.70}$$

where the damaged strain concentration factor $\bar{\mathbf{A}}^{\mathrm{R}}$ is given by:

$$\bar{A}^{\mathrm{R}}_{ijkl} = M^{\mathrm{R}}_{mnij} A^{\mathrm{R}}_{mnpq} M^{-1}_{pqkl}. \tag{1.71}$$

One then assumes the generalized Hooke's law to hold for each of the phases in the configuration $\bar{C}$, that is

$$\bar{\sigma}^{\mathrm{R}}_{ij} = E^{\mathrm{R}}_{ijkl} \bar{\varepsilon}^{\mathrm{R}}_{kl}, \tag{1.72}$$

where E^{R}_{ijkl} is the constant elasticity tensor for the phase material. Substituting Equations (1.63) and (1.72) into Equation (1.36) along with Equations (1.44) and (1.45) and simplifying, one obtains the following local-overall relation for the undamaged elasticity tensors:

$$E_{ijkl} = c^{\mathrm{M}} A^{\mathrm{M}}_{ijmn} E^{\mathrm{M}}_{mnkl} + c^{\mathrm{F}} A^{\mathrm{F}}_{ijmn} E^{\mathrm{F}}_{mnkl}. \tag{1.73}$$

Assuming similar local relations to hold as those of Equation (1.72) in the configuration C, one has

$$\sigma^{\mathrm{R}}_{ij} = \bar{E}^{\mathrm{R}}_{ijkl} \varepsilon^{\mathrm{R}}_{kl}, \tag{1.74}$$

where the tensor $\bar{E}^{\mathrm{R}}_{ijkl}$ is the damaged elasticity tensor for the phase material. Substituting Equations (1.66) and (1.74) into an Equation similar to Equation (1.36), written in the configuration C, along with Equations (1.69) and simplifying, one obtains

$$\bar{E}_{ijkl} = \bar{c}^{\mathrm{M}} \bar{A}^{\mathrm{M}}_{ijmn} \bar{E}^{\mathrm{M}}_{mnkl} + \bar{c}^{\mathrm{F}} \bar{A}^{\mathrm{F}}_{ijmn} \bar{E}^{\mathrm{F}}_{mnkl}. \tag{1.75}$$

Next, one considers the transformation Equations for the local moduli of elasticity E^{M}_{ijkl} and E^{F}_{ijkl}. Starting with Equation (1.75) and substituting for $\bar{\mathbf{E}}$ from Equation (1.67), for $\bar{\mathbf{A}}^{\mathrm{M}}$ and $\bar{\mathbf{A}}^{\mathrm{F}}$ from Equations (1.71) and for $\mathbf{E}$ from Equation (1.73) and simplifying, one obtains:

$$\bar{E}^{\mathrm{R}}_{ijkl} = \frac{c^{\mathrm{R}}}{\bar{c}^{\mathrm{R}}} M^{-1}_{mnij} E^{\mathrm{R}}_{mnpq} \left(M^{\mathrm{R}}_{pqkl} \right)^{-1}. \tag{1.76}$$

The remainder of this Section is left to determine a proper transformation relation for the phase volume fractions c^{M} and c^{F}. The authors see no direct way of deriving such Equations at the present time. However, in view of the relation given in Equation (1.67), the scalar ratios $c^{\mathrm{M}}/\bar{c}^{\mathrm{M}}$ and $c^{\mathrm{F}}/\bar{c}^{\mathrm{F}}$ can be determined by comparison. Therefore, the following transformation Equation is listed here without proof.

$$c^{\mathrm{R}} = \frac{\bar{c}^{\mathrm{R}}}{9} \left(M^{\mathrm{R}}_{pqij} \right)^{-1} M_{pqij}. \tag{1.77}$$

Upon substituting the above transformation Equation for c^{R} into Equation (1.76), one obtains simple relations for the transformation of the local moduli of elasticity similar to that of Equation (1.67). It should be noted that Equation (1.77) does not imply a change in the phase volume fractions. The quantities

$\bar{c}^{\mathrm{M}}$ and $\bar{c}^{\mathrm{F}}$ represent effective phase volume fractions in the fictitious undamaged configurations $\bar{C}^{\mathrm{M}}$ and $\bar{C}^{\mathrm{F}}$, respectively. They should be regarded similarly to the effective stresses $\bar{\sigma}^{\mathrm{M}}$ and $\bar{\sigma}^{\mathrm{F}}$ where they do not represent actual quantities, but effective quantities in the context of continuum damage mechanics. The constitutive theory as well as the relevant transformation Equations have been presented for the analysis of damage and small elastic deformation of fiber-reinforced composite materials. This is illustrated in detail in Section 1.4.6 for the case of uniaxial tension.

The rest of this Section is left for a brief discussion of the stress and strain concentration factors. The simplest model available involves the Voigt assumption where the matrix, fiber and overall strain rates are assumed equal. In our case, the Voigt model is applied to the configuration $\bar{C}$ in the form $\bar{\varepsilon}_{ij}^{\mathrm{M}} = \bar{\varepsilon}_{ij}^{\mathrm{F}} = \bar{\varepsilon}_{ij}$. Substituting these into Equation (1.44) immediately results in $A_{ijkl}^{\mathrm{M}} = A_{ijkl}^{\mathrm{F}} = \frac{1}{2}\left(\delta_{ik}\delta_{jl} + \delta_{il}\delta_{jk}\right)$. Using this result along with Equations (1.35), (1.63) and (1.72) will yield $B_{ijkl}^{\mathrm{M}} = E_{ijmn}^{\mathrm{M}} E_{klmn}^{-1}$ and $B_{ijkl}^{\mathrm{F}} = E_{ijmn}^{\mathrm{F}} E_{klmn}^{-1}$. Another widely known model is the Vanishing Fiber Diameter (VFD) model. In this model, it is assumed that the cylindrical fibers have vanishing diameters while occupying a finite volume fraction of the composite. However, the resulting Equations are not as simple as those of the Voigt model and the reader is referred to Dvorak and Bahei-El-Din (1979, 1982) for a detailed discussion of the VFD model. Finally, it should be noted that more sophisticated models (Mori and Tanaka, 1973) can also be used.

1.4.5
Evolution of Damage

In order to study the evolution of damage in composite materials, one first needs to investigate the nature of the fourth-rank damage effect tensor **M**. It has been shown (Voyiadjis and Kattan, 1992) that using the Voigt notation for stresses and strains (i.e., representing them as vectors instead of tensors), the tensor M_{ijkl} can be represented by a 6×6 matrix in terms of a second-rank damage tensor ϕ. Therefore, it is clear (Voyiadjis and Kattan, 1992) that the study of damage evolution involves the determination of an appropriate kinetic Equation for the tensor ϕ_{ij}. One introduces the generalized thermodynamic force y_{ij} that is associated with ϕ_{ij} by the definition (Lemaitre, 1985):

$$y_{ij} = \frac{\partial U}{\partial \phi_{ij}}, \tag{1.78}$$

such that $\dot{\phi}_{ij} y_{ij}$ is the power dissipated due to the damage. The criterion for damage evolution used here is that proposed by Lee *et al.* (1985) and is given by the function $g(\mathbf{y}, B)$ defined by:

$$g(\mathbf{y}, B) = \frac{1}{2} J_{ijkl} y_{ij} y_{kl} - \mathrm{B}(\beta) = 0, \tag{1.79}$$

where $B(\beta)$ is a function of the overall damage parameter β and J_{ijkl} is a constant fourth-rank tensor that can be represented by a constant 6×6 matrix (Lee, et al., 1985; Voyiadjis and Kattan, 1992). In order to develop an evolution Equation for the damage variable ϕ, one considers the power of dissipation Π given by:

$$\Pi = -y_{pq} \dot{\phi}_{pq} - B\dot{\beta}. \tag{1.80}$$

The problem is to extremize Π subject to the constraint $dg = 0$. Therefore, one introduces the Lagrange multiplier $\dot{\lambda}$ and uses the Lagrange multiplier method to obtain $\dot{\lambda} = \dot{\beta}$ and

$$\dot{\phi}_{ij} = -\dot{\beta} \frac{\partial g}{\partial y_{ij}}. \tag{1.81}$$

In order to determine $\dot{\beta}$, one uses the consistency condition $dg = 0$ in the form

$$\frac{\partial g}{\partial y_{pq}} \dot{y}_{pq} + \frac{\partial g}{\partial B} \dot{B} = 0. \tag{1.82}$$

Substituting for the partial derivatives of g from Equation (1.79) into Equation (1.82) and solving for $\dot{\beta}$, one obtains:

$$\dot{\beta} = \dot{\lambda} = \frac{J_{pqmn} y_{mn} \dot{y}_{pq}}{\partial B / \partial \beta}. \tag{1.83}$$

Substituting the above expression of $\dot{\beta}$ into Equation (1.81), one obtains the required evolution Equation for the damage tensor ϕ_{ij}:

$$\dot{\phi}_{ij} = \frac{-J_{pqmn} \dot{y}_{mn} y_{pq}}{\partial B / \partial \beta} \frac{\partial g}{\partial y_{ij}}. \tag{1.84}$$

The solution of the above kinetic Equation hinges on the determination of an appropriate expression for the function $B(\beta)$. One may use a linear function in the form $B(\beta) = c_1 \beta + c_2$ where c_1 and c_2 are constants. This is motivated by analogy to the isotropic hardening parameter κ in the theory of plasticity, where the evolution of κ is taken to be $\dot{\kappa} = \left(\dot{\varepsilon}''_{ij} \dot{\varepsilon}''_{ij} \right)^{1/2}$ where $\dot{\varepsilon}''$ is the rate of plastic strain. Analogously, using $\dot{B} = \left(\dot{\beta} \dot{\beta} \right)^{1/2}$ will yield a linear function $B(\beta)$ as proposed above. An example is given in the next Section where the evolution Equation is solved for the case of uniaxial tension with a linear function $B(\beta)$.

1.4.6
Example 1: Uniaxial Tension of a Unidirectional Lamina

Consider a unidirectional fiber-reinforced thin lamina that is subjected to a uniaxial tensile force T along the x_1-direction. The lamina is made of an elastic matrix with elastic fibers aligned along the x_1-direction. The x_2- and x_3-axes are assumed to lie in the plane of the lamina. Let $\mathrm{d}S$ be the cross-sectional area of the lamina with $\mathrm{d}S^{\mathrm{M}}$ and $\mathrm{d}S^{\mathrm{F}}$ being the cross-sectional areas of the matrix and fibers, respectively. In the fictitious undamaged configuration, let the cross-sectional areas of the lamina, matrix and fibers be denoted by $\mathrm{d}\bar{S}$, $\mathrm{d}\bar{S}^{\mathrm{M}}$ and $\mathrm{d}\bar{S}^{\mathrm{F}}$, respectively. Since the lamina strictly consists of a matrix and fibers, it is clear that $\mathrm{d}S^{\mathrm{M}} + \mathrm{d}S^{\mathrm{F}} = \mathrm{d}S$, $\mathrm{d}\bar{S}^{\mathrm{M}} + \mathrm{d}\bar{S}^{\mathrm{F}} = \mathrm{d}\bar{S}$, $\mathrm{d}\bar{S} \leq \mathrm{d}S$, $\mathrm{d}\bar{S}^{\mathrm{M}} \leq \mathrm{d}S^{M}$ and $\mathrm{d}\bar{S}^{\mathrm{F}} \leq \mathrm{d}S^{F}$ (Kattan and Voyiadjis, 1992).

The overall stress, strain and damage tensors σ, ε, and ϕ for this problem can be represented using the following vectors

$$\sigma \equiv \begin{Bmatrix} \sigma \\ 0 \\ 0 \end{Bmatrix}, \quad \varepsilon \equiv \begin{Bmatrix} \varepsilon_1 \\ \varepsilon_2 \\ \varepsilon_3 \end{Bmatrix}, \quad \phi \equiv \begin{Bmatrix} \phi_1 \\ \phi_2 \\ \phi_3 \end{Bmatrix}, \tag{1.85}$$

with similar vector representations for their corresponding effective and local counterparts. The uniaxial stress σ appearing in Equation (1.85) is clearly given by $\sigma = T/\mathrm{d}S$ with the uniaxial effective stress given by $\bar{\sigma} = T/\mathrm{d}\bar{S}$. The overall damage variable ϕ_1 is defined by (Kachanov, 1958)

$$\phi_1 = \frac{\mathrm{d}S\mathrm{d}\bar{S}}{\mathrm{d}S}. \tag{1.86}$$

It is clear from Equation (1.86) that ϕ_1 takes the values between 0 for undamaged material to 1 for (theoretically) complete rupture. However, the actual value ϕ_{cr} where failure occurs is less than 1 and satisfies $0 \leq \phi < \phi_{\mathrm{cr}} < 1$. Two local damage variables ϕ_1^{M} and ϕ_1^{F} can be analogously introduced and defined by:

$$\phi_1^{\mathrm{R}} = \frac{\mathrm{d}S^{\mathrm{R}} - \mathrm{d}\bar{S}^{\mathrm{R}}}{\mathrm{d}S^{\mathrm{R}}}. \tag{1.87}$$

It follows directly from Equation (1.87) that $0 \leq \phi_1^{\mathrm{R}} \leq 1$. Using Equations (1.86) and (1.87) along with the area relations discussed in the beginning of this Section, one can easily derive the following relation between the local and overall damage variables:

$$\phi_1 = c^{\mathrm{M}}\phi_1^{M} + c^{\mathrm{F}}\phi_1^{\mathrm{F}}. \tag{1.88}$$

It should be mentioned that the uniaxial local and overall stresses σ, σ^{M} and σ^{F} satisfy a similar relation to that of Equation (1.88) and is given in tensor form by Equation (1.36). The relation between the overall stress σ and its effective counterpart $\bar{\sigma}$ can be easily shown to be given by

$$\bar{\sigma} = \frac{\sigma}{1 - \phi_1}. \tag{1.89}$$

Using Equation (1.36) and a similar Equation for the effective stress, one can assume the local stresses to be given by:

$$\bar{\sigma}^R = \frac{\sigma^R}{1 - \phi_1^R}. \tag{1.90}$$

In view of Equation (1.89), it is clear that the relations (1.90) satisfy the requirements given by Equation (1.36). Comparing Equations (1.89) and (1.90) with the general transformation Equations (1.47) and (1.57) and considering the notation of Equation (1.85) for this problem, the damage effect tensors **M**, $\mathbf{M}^M$ and $\mathbf{M}^F$ can be represented by the following matrices:

$$\mathbf{M} \equiv \begin{bmatrix} \frac{1}{1-\phi_1} & 0 & 0 \\ 0 & \frac{1}{1-\phi_2} & 0 \\ 0 & 0 & \frac{1}{1-\phi_3} \end{bmatrix}, \tag{1.91}$$

$$\mathbf{M}^R \equiv \begin{bmatrix} \frac{1}{1-\phi_1^R} & 0 & 0 \\ 0 & \frac{1}{1-\phi_2^R} & 0 \\ 0 & 0 & \frac{1}{1-\phi_3^R} \end{bmatrix}, \tag{1.92}$$

It should be mentioned that the matrix representation of the damage effect tensor **M** of Equation (1.91) applies only to the problem of uniaxial tension considered here. For a general matrix representation of the tensor **M**, the reader is referred to the recent paper by Voyiadjis and Kattan (1992).

The overall elasticity tensor E_{ijkl} can be represented by the following matrix where an orthotropic material is assumed:

$$\mathbf{E}^{-1} \equiv \begin{bmatrix} \frac{1}{E_1} & \frac{-\nu_{21}}{E_2} & \frac{-\nu_{31}}{E_3} \\ \frac{-\nu_{12}}{E_1} & \frac{1}{E_2} & \frac{-\nu_{32}}{E_3} \\ \frac{-\nu_{13}}{E_1} & \frac{-\nu_{23}}{E_2} & \frac{1}{E_3} \end{bmatrix}. \tag{1.93}$$

Using the representations of **M** and **E** of Equations (1.91) and (1.93), and substituting them into the transformation Equation (1.67), one obtains the following matrix for the damaged elasticity tensor $\bar{E}_{ijkl}$:

$$\bar{\mathbf{E}}^{-1} \equiv \begin{bmatrix} \frac{1}{E_1(1-\phi_1)^2} & \frac{-\nu_{21}}{E_2(1-\phi_1)(1-\phi_2)} & \frac{-\nu_{31}}{E_3(1-\phi_1)(1-\phi_3)} \\ \frac{-\nu_{12}}{E_1(1-\phi_1)(1-\phi_2)} & \frac{1}{E_2(1-\phi_2)^2} & \frac{-\nu_{32}}{E_3(1-\phi_2)(1-\phi_3)} \\ \frac{-\nu_{13}}{E_1(1-\phi_1)(1-\phi_3)} & \frac{-\nu_{23}}{E_2(1-\phi_2)(1-\phi_3)} & \frac{1}{E_3(1-\phi_3)^2} \end{bmatrix}. \tag{1.94}$$

Considering a matrix representation for $\bar{\mathbf{E}}^{-1}$ similar to that of Equation (1.93) but with all quantities replaced by barred quantities and comparing it with the matrix in Equation (1.94), one obtains the following transformation Equations for the overall elastic properties:

$$\bar{E}_i = E_i(1-\phi_i)^2, \quad i = 1, 2, 3 \text{ (no sum)}, \tag{1.95}$$

$$\bar{\nu}_{ij} = \nu_{ij}\frac{1-\phi_i}{1-\phi_j}, \quad i,j = 1, 2, 3 \text{ (no sum)}. \tag{1.96}$$

Next, one uses the transformation Equation (1.60) for the phase stress concentration factors, and substitutes for the damage effect tensors from Equations (1.91) and (1.92) to derive the following matrix representation for the damaged phase stress concentration factor $\bar{B}^R_{ijkl}$:

$$\bar{\mathbf{B}}^{\mathrm{R}} \equiv \begin{bmatrix} \frac{1-\phi^{\mathrm{R}}_1}{1-\phi_1}B^{\mathrm{R}}_{11} & \frac{1-\phi^{\mathrm{R}}_1}{1-\phi_2}B^{\mathrm{R}}_{12} & \frac{1-\phi^{\mathrm{R}}_1}{1-\phi_3}B^{\mathrm{R}}_{13} \\ \frac{1-\phi^{\mathrm{R}}_2}{1-\phi_1}B^{\mathrm{R}}_{21} & \frac{1-\phi^{\mathrm{R}}_2}{1-\phi_2}B^{\mathrm{R}}_{22} & \frac{1-\phi^{\mathrm{R}}_2}{1-\phi_3}B^{\mathrm{R}}_{23} \\ \frac{1-\phi^{\mathrm{R}}_3}{1-\phi_1}B^{\mathrm{R}}_{31} & \frac{1-\phi^{\mathrm{R}}_3}{1-\phi_2}B^{\mathrm{R}}_{32} & \frac{1-\phi^{\mathrm{R}}_3}{1-\phi_3}B^{\mathrm{R}}_{33} \end{bmatrix}, \tag{1.97}$$

where the terms B^R_{ij} are the elements of the matrix representation of B^R_{ijkl}.

Similarly, one uses the transformation Equation (1.71) for the strain concentration factors to derive the following matrix representation for the damaged phase strain concentration factor $\bar{A}^R_{ijkl}$:

$$\bar{\mathbf{A}}^{\mathrm{R}} \equiv \begin{bmatrix} \frac{1-\phi_1}{1-\phi^{\mathrm{R}}_1}A^{\mathrm{R}}_{11} & \frac{1-\phi_1}{1-\phi^{\mathrm{R}}_2}A^{\mathrm{R}}_{12} & \frac{1-\phi_1}{1-\phi^{\mathrm{R}}_3}A^{\mathrm{R}}_{13} \\ \frac{1-\phi_2}{1-\phi^{\mathrm{R}}_1}A^{\mathrm{R}}_{21} & \frac{1-\phi_2}{1-\phi^{\mathrm{R}}_2}A^{\mathrm{R}}_{22} & \frac{1-\phi_2}{1-\phi^{\mathrm{R}}_3}A^{\mathrm{R}}_{23} \\ \frac{1-\phi_3}{1-\phi^{\mathrm{R}}_1}A^{\mathrm{R}}_{31} & \frac{1-\phi_3}{1-\phi^{\mathrm{R}}_2}A^{\mathrm{R}}_{32} & \frac{1-\phi_3}{1-\phi^{\mathrm{R}}_3}A^{\mathrm{R}}_{33} \end{bmatrix}, \tag{1.98}$$

where the terms A^R_{ij} are the elements in the matrix representation of A^R_{ijkl}.

Finally, one writes the transformation Equations for the volume fractions c^{M} and c^{F}. Using Equation (1.77) along with the matrix representations (1.91) and (1.92), one derives:

$$\bar{c}^{\mathrm{R}} = c^{\mathrm{R}}\frac{1-\phi^{\mathrm{R}}_1}{1-\phi_1}. \tag{1.99}$$

Alternatively, the above relations can be derived independently using the definitions of c^{M} and c^{F} as area fractions for this problem, along with Equations (1.86) and (1.87). Finally, one can use Equations (1.76) and (1.77) to derive transformation Equations for the local elastic properties. However, the resulting Equations are similar to Equations (1.95) and (1.96) with superscripts "M" or "F" and will not be listed here.

In order to characterize damage evolution for this problem, one uses Equation (1.84). For this problem, the kinetic Equation (1.84) reduces to

$$\dot{\phi}_1 = \frac{-y_1^2 \dot{y}_1}{\partial B / \partial \beta}, \tag{1.100}$$

where ϕ_1 and y_1 stand for the tensor components ϕ_{11} and y_{11}, respectively. Using a linear function $B(\beta) = c_1 \beta + c_2$ where c_1 and c_2 are constants, substituting it into Equation (1.100) and solving the differential Equation, one obtains

$$\phi_1 = -\frac{-y_1^3}{3c_1}. \tag{1.101}$$

The above Equation represents the relation between the damage variable ϕ_1 and its associated thermodynamic generalized force y_1 for the case of uniaxial tension. One then substitutes for y_1 from Equation (1.78) along with Equation (1.64) into Equation (1.101) to obtain:

$$\frac{\phi_1}{(1-\phi_1)^6} = \frac{E^3}{3c_1} \varepsilon_1^6. \tag{1.102}$$

Equation (1.102) represents the overall strain-damage relation for the case of uniaxial tension. Similar relations can be derived for the local strain and damage variables.

1.4.7 Example 2: A Unidirectional Lamina Under Plane Stress

Consider a unidirectional fiber-reinforced thin lamina that is subjected to a case of plane stress in the 1-2 plane. The lamina is made of an elastic material with elastic fibers aligned along the x_1-axis. Both the x_1- and x_2-axes are assumed to lie in the plane of the lamina, while the x_3-axis lies in the transverse direction to that plane. A complete damage state is considered, although the lamina is under plane stress. Therefore, all the damage variables are assumed nonzero in this example. For this case of plane stress, the stress and damage tensors are represented by the following matrices:

$$[\sigma_{ij}] = \begin{bmatrix} \sigma_{11} & \sigma_{12} & 0 \\ \sigma_{12} & \sigma_{22} & 0 \\ 0 & 0 & 0 \end{bmatrix}, \tag{1.103}$$

$$[\phi_{ij}] = \begin{bmatrix} \phi_{11} & \phi_{12} & \phi_{13} \\ \phi_{21} & \phi_{22} & \phi_{23} \\ \phi_{31} & \phi_{32} & \phi_{33} \end{bmatrix}. \tag{1.104}$$

The transformation Equation (1.47) gives rise to a nonsymmetric effective stress tensor. Therefore, before proceeding with this example, one needs to symmetrize the effective stress tensor $\bar{\sigma}$. One of the most popular symmetrization procedures is given by the Equation

$$\bar{\sigma}_{ij} = \frac{1}{2}\left[\sigma_{ik}\left(\delta_{kj} - \phi_{kj}\right)^{-1} + \left(\delta_{il} - \phi_{il}\right)^{-1}\sigma_{lj}\right]. \tag{1.105}$$

Before substituting Equations (1.103) and (1.104) into Equation (1.105), one needs to find the inverse of the tensor $\delta_{ij} - \phi_{ij}$. This is found simply by obtaining the inverse of the matrix $\left[\delta_{ij} - \phi_{ij}\right]$ through the use of the symbolic manipulation program REDUCE. The resulting matrix is given as:

$$\left[\delta_{ij} - \phi_{ij}\right]^{-1} = \frac{1}{\Delta} \times \begin{bmatrix} (1-\phi_{22})(1-\phi_{33}) - \phi_{23}^2 & \phi_{13}\phi_{23} + \phi_{12}(1-\phi_{33}) & \phi_{12}\phi_{23} + \phi_{13}(1-\phi_{22}) \\ \phi_{13}\phi_{23} + \phi_{12}(1-\phi_{33}) & (1-\phi_{11})(1-\phi_{33}) - \phi_{13}^2 & \phi_{12}\phi_{13} + \phi_{23}(1-\phi_{11}) \\ \phi_{12}\phi_{23} + \phi_{13}(1-\phi_{22}) & \phi_{12}\phi_{13} + \phi_{23}(1-\phi_{11}) & (1-\phi_{11})(1-\phi_{22}) - \phi_{12}^2 \end{bmatrix}, \tag{1.106}$$

where Δ is given by

$$\Delta = (1-\phi_{11})(1-\phi_{22})(1-\phi_{33}) - \phi_{23}^2(1-\phi_{11}) - \phi_{13}^2(1-\phi_{22}) - \phi_{12}^2(1-\phi_{33}) - 2\phi_{12}\phi_{23}\phi_{13}. \tag{1.107}$$

Next, one substitutes Equations (1.103) and (1.106) into Equation (1.105) and simplifies the resulting matrix. Using the vector representation $\left[\sigma_{11} \quad \sigma_{22} \quad \sigma_{12}\right]^T$ for the stress tensor σ, and likewise using $\left[\bar{\sigma}_{11} \quad \bar{\sigma}_{22} \quad \bar{\sigma}_{12}\right]^T$ for the effective stress tensor $\bar{\sigma}$, the resulting Equation can be rewritten as:

$$\begin{Bmatrix} \bar{\sigma}_{11} \\ \bar{\sigma}_{22} \\ \bar{\sigma}_{12} \end{Bmatrix} = \begin{bmatrix} M_{11} & M_{12} & M_{13} \\ M_{21} & M_{22} & M_{23} \\ M_{31} & M_{32} & M_{33} \end{bmatrix} \begin{Bmatrix} \sigma_{11} \\ \sigma_{22} \\ \sigma_{12} \end{Bmatrix}, \tag{1.108}$$

where the coefficients of the matrix [**M**] of Equation (1.108) are given by:

$$M_{11} = \frac{(1-\phi_{22})(1-\phi_{33}) - \phi_{23}^2}{\Delta}, \tag{1.109a}$$

$$M_{22} = \frac{(1-\phi_{11})(1-\phi_{33}) - \phi_{13}^2}{\Delta}, \tag{1.109b}$$

$$M_{33} = \frac{M_{11} + M_{22}}{2}, \tag{1.109c}$$

$$M_{12} = M_{21} = 0, \tag{1.109d}$$

$$M_{13} = 2M_{31} = \frac{\phi_{12}\phi_{23} + \phi_{12}(1-\phi_{33})}{\Delta}, \tag{1.109e}$$

$$M_{23} = 2M_{32} = M_{13}. \tag{1.109f}$$

The above Equations were obtained using the symbolic manipulation program REDUCE.

For the case of plane stress discussed in this Section, the overall elasticity tensor E_{ijkl} is written here for an orthotropic material indirectly in the following form:

$$\mathbf{E}^{-1} \equiv \begin{bmatrix} \frac{1}{E_1} & \frac{-\nu_{12}}{E_1} & 0 \\ \frac{-\nu_{21}}{E_2} & \frac{1}{E_2} & 0 \\ 0 & 0 & \frac{1}{G_{12}} \end{bmatrix}. \tag{1.110}$$

Using the representation of **M** and $\mathbf{E}^{-1}$ in Equations (1.109) and (1.110) and substituting them into the transformation Equation (1.67), one obtains the following matrix for the damaged elasticity tensor $\bar{E}_{ijkl}$:

$$\bar{\mathbf{E}}^{-1} \equiv \begin{bmatrix} E_{11} & E_{12} & E_{13} \\ E_{12} & E_{22} & E_{23} \\ E_{13} & E_{23} & E_{33} \end{bmatrix}, \tag{1.111}$$

where the terms in the above matrix are given by:

$$E_{11} = \frac{M_{11}^2}{E_1} + \frac{M_{13}^2}{2G_{12}}, \tag{1.112a}$$

$$E_{22} = \frac{M_{22}^2}{E_2} + \frac{M_{13}^2}{2G_{12}}, \tag{1.112b}$$

$$E_{33} = \frac{1}{2}M_{13}^2\left(\frac{1}{E_1} + \frac{1}{E_2} - \frac{\nu_{12}}{E_1} - \frac{\nu_{21}}{E_2}\right) + \frac{(M_{11}+M_{22})^2}{4G_{12}}, \tag{1.112c}$$

$$E_{12} = -\nu_{12}\frac{M_{11}+M_{22}}{E_1} + \frac{M_{13}^2}{2G_{12}}, \tag{1.112d}$$

$$E_{13} = M_{13}\left[\frac{M_{11}}{E_1} - \nu_{12}\frac{M_{11}}{E_1} + \frac{(M_{11}+M_{22})}{2G_{12}}\right], \tag{1.112e}$$

$$E_{23} = M_{13}\left[\frac{M_{22}}{E_2} - \nu_{21}\frac{M_{22}}{E_2} + \frac{(M_{11}+M_{22})}{2G_{12}}\right]. \tag{1.112f}$$

The inverse of the matrix in Equation (1.111) should be equivalent to the effective elasticity tensor $\bar{E}_{ijkl}^{-1}$ given by:

$$\bar{\mathbf{E}}^{-1} \equiv \begin{bmatrix} \frac{1}{\bar{E}_1} & -\frac{\bar{\nu}_{12}}{\bar{E}_1} & 0 \\ -\frac{\bar{\nu}_{21}}{\bar{E}_2} & \frac{1}{\bar{E}_2} & 0 \\ 0 & 0 & \frac{1}{\bar{G}_{12}} \end{bmatrix}. \tag{1.113}$$

The relations between the elastic constant E_1, E_2, ν_{12}, G_{12} and the effective elastic variables $\bar{E}_1$, $\bar{E}_2$, $\bar{\nu}_{12}$, and $\bar{G}_{12}$ can be obtained by equating Equations (1.111) and (1.113). After some lengthy algebraic manipulations one arrives at the following relations:

$$\bar{E}_1 = \frac{E_1}{M_{11}}\left[\frac{M_{11}+M_{22}}{M_{11}^2 + M_{11}M_{22} - (1-\nu_{12})M_{13}^2}\right], \tag{1.114a}$$

$$\bar{E}_2 = \frac{E_2}{M_{22}}\left[\frac{M_{11}+M_{22}}{M_{22}^2 + M_{11}M_{22} - (1-\nu_{21})M_{13}^2}\right], \tag{1.114b}$$

$$\bar{\nu}_{12} = \frac{\nu_{12}\left(M_{22}^2 + M_{11}M_{22}\right) - (1-\nu_{12})M_{13}^2}{M_{11}^2 + M_{11}M_{22} - (1-\nu_{12})M_{13}^2}, \tag{1.114c}$$

$$\bar{G}_{12} = G_{12}\frac{4M_{11}M_{22}}{(M_{11}+M_{22})^2\left(M_{11}M_{22} - M_{13}^2\right)}. \tag{1.114d}$$

In addition, one obtains the following relation:

$$\nu_{12}E_2 = \nu_{21}E_1. \tag{1.115}$$

The above relation holds for the damaged composite system and is similar to the usual composite relation for effective quantities $\bar{\nu}_{12}\bar{E}_2 = \bar{\nu}_{21}\bar{E}_1$. It should be noted that in Equation (1.114a), $\bar{E}_1$ is a function of both E_1 and ν_{12} in addition to the damage variables. Similarly, in Equation (1.114b), $\bar{E}_2$ depends on E_2, ν_{21} and the damage variables. On the other hand, it is clear from Equations (1.114c) and (1.114d) that $\bar{\nu}_{12}$ depends only on ν_{12} and $\bar{G}_{12}$ depends only on G_{12} in addition to the damage variables. An expression for $\bar{\nu}_{12}$ in terms of ν_{12} can be obtained similar to Equation (1.114c).

When using the principal damage variables ϕ_1, ϕ_2, and ϕ_3, Equations (1.114) reduce to:

$$\bar{E}_1 = E_1\left(\frac{1}{M_{11}}\right)^2, \tag{1.116a}$$

$$\bar{E}_2 = E_2\left(\frac{1}{M_{22}}\right)^2, \tag{1.116b}$$

$$\bar{\nu}_{12} = \nu_{12}\frac{M_{22}}{M_{11}}, \tag{1.116c}$$

$$\bar{G}_{12} = G_{12}\left(\frac{2}{M_{11} + M_{22}}\right)^2. \tag{1.116d}$$

Substituting for M_{11}, M_{22}, and Δ from Equations (1.109) into Equations (1.116), keeping in mind that $\phi_{13} = \phi_{23} = \phi_{12} = M_{13} = 0$, one obtains:

$$\bar{E}_1 = E_1(1 - \phi_1)^2, \tag{1.117a}$$

$$\bar{E}_2 = E_2(1 - \phi_2)^2, \tag{1.117b}$$

$$\bar{\nu}_{12} = \nu_{12}\frac{1 - \phi_1}{1 - \phi_2}, \tag{1.117c}$$

$$\bar{G}_{12} = G_{12}\left[\frac{2(1 - \phi_1)^2(1 - \phi_2)^2(1 - \phi_3)^3}{(1 - \phi_1) + (1 - \phi_2)^2}\right]^2. \tag{1.117d}$$

Equations (1.117a) to (1.117c) are similar to Equations (1.95) and (1.96). Thus, the expressions involving principal damage variables reduce to those of the uniaxial tension case of the previous example. The only new Equation here is Equation (1.117d) which relates shear quantities that do not exist in the uniaxial tension case.

The phase damage effect tensor M^R_{ijkl} is considered here to have the same form as the overall damage effect tensor **M** given in Equations (1.108) and (1.109). This tensor takes the form:

$$\left[M^R_{ijkl}\right] = \frac{1}{\Delta^R}\begin{bmatrix} M^R_{11} & 0 & M^R_{13} \\ 0 & M^R_{22} & M^R_{13} \\ \frac{1}{2}M^R_{13} & \frac{1}{2}M^R_{13} & \frac{1}{2}\left(M^R_{11}+M^R_{22}\right) \end{bmatrix}, \tag{1.118}$$

where Δ^R, M^R, M^R_{22} and M^R_{13} are given by:

$$\begin{aligned}\Delta^R = {} & \left(1-\phi^R_{11}\right)\left(1-\phi^R_{22}\right)\left(1-\phi^R_{33}\right) - \left(\phi^R_{23}\right)^2\left(1-\phi^R_{11}\right) \\ & - \left(\phi^R_{13}\right)^2\left(1-\phi^R_{22}\right) - \left(\phi^R_{12}\right)^2\left(1-\phi^R_{33}\right)2\phi^R_{12}\phi^R_{23}\phi^R_{13},\end{aligned} \tag{1.119a}$$

$$M^R_{11} = \frac{\left(1-\phi^R_{22}\right)\left(1-\phi^R_{33}\right) - \left(\phi^R_{23}\right)^2}{\Delta}, \tag{1.119b}$$

$$M^R_{22} = \frac{\left(1-\phi^R_{11}\right)\left(1-\phi^R_{33}\right) - \left(\phi^R_{13}\right)^2}{\Delta}, \tag{1.119c}$$

$$M^R_{13} = \frac{\phi^R_{12}\phi^R_{23} + \phi^R_{12}\left(1-\phi^R_{33}\right)}{\Delta}. \tag{1.119d}$$

The expressions of **M** and $\mathbf{M}^R$ given in Equations (1.109) and (1.119), respectively, are substituted into the transformation Equation (1.60) to obtain the matrix and fiber stress concentration factors for the case of plane stress. The phase stress concentration factors $\bar{B}^R_{ij}$ are then given by:

$$\begin{Bmatrix} \bar{B}^R_{11} \\ \bar{B}^R_{22} \\ \bar{B}^R_{33} \\ \bar{B}^R_{12} \\ \bar{B}^R_{13} \\ \bar{B}^R_{23} \end{Bmatrix} = L\begin{bmatrix} l_{11} & l_{12} & l_{13} & l_{14} & l_{15} & l_{16} \\ l_{12} & l_{22} & l_{23} & l_{24} & l_{25} & l_{26} \\ l_{13} & l_{23} & l_{33} & l_{34} & l_{35} & l_{36} \\ l_{41} & l_{42} & l_{43} & l_{44} & l_{45} & l_{46} \\ l_{51} & l_{52} & l_{53} & l_{54} & l_{55} & l_{56} \\ l_{61} & l_{62} & l_{63} & l_{64} & l_{65} & l_{66} \end{bmatrix}\begin{Bmatrix} B^R_{11} \\ B^R_{22} \\ B^R_{33} \\ B^R_{12} \\ B^R_{13} \\ B^R_{23} \end{Bmatrix}, \tag{1.120}$$

where L is given by:

$$L = \frac{\Delta}{\Delta^R\left[M^R_{11}\left(M^R_{22}\right) + \left(M^R_{11}\right)^2 M^R_{22} - \left(M^R_{13}\right)^2\left(M^R_{11}+M^R_{22}\right)\right]}. \tag{1.121}$$

The coefficients l_{ij} in the above matrix are given in Equations (1.122–1.154) and the terms B^R_{ij} are the elements of the matrix representation of the tensor B^R_{ijkl}.

In the same way, one can use the transformation Equation (1.71) for the strain concentration factors $\bar{A}^M_{ijkl}$ and $\bar{A}^F_{ijkl}$ to obtain results corresponding to

Equation (1.120). The resulting coefficients look similar to those in Equations (1.122) to (1.154) and are not shown here for the sake of simplicity.

$$l_{11} = M_{11}\left[\left(M^{R}_{22}\right)^2 + M^2_{11}M^2_{22} - \left(M^{R}_{13}\right)^2\right], \tag{1.122}$$

$$l_{12} = 0, \tag{1.123}$$

$$l_{13} = -M_{11}M^{R}_{22}M^{R}_{13}, \tag{1.124}$$

$$l_{14} = -M_{11}\left(M^{R}_{22}\right)^2, \tag{1.125}$$

$$l_{15} = M_{13}\left[\left(M^{R}_{22}\right)^2 + M^2_{11}M^2_{22} - \left(M^{R}_{13}\right)^2\right] - 2M_{11}M^{R}_{22}M^{R}_{13}, \tag{1.126}$$

$$l_{16} = -M_{13}\left(M^{R}_{13}\right)^2, \tag{1.127}$$

$$l_{22} = M_{22}\left[\left(M^{R}_{11}\right)^2 + M^2_{11}M^2_{22} - \left(M^{R}_{13}\right)^2\right], \tag{1.128}$$

$$l_{23} = -M_{13}M^{R}_{11}M^{R}_{13}, \tag{1.129}$$

$$l_{24} = M_{22}\left(M^{R}_{13}\right)^2, \tag{1.130}$$

$$l_{25} = M_{13}\left(M^{R}_{13}\right)^2, \tag{1.131}$$

$$l_{26} = M_{13}\left[\left(M^{R}_{11}\right)^2 + M^2_{11}M^2_{22} - \left(M^{R}_{13}\right)^2\right] - 2M_{22}M^{R}_{11}M^{R}_{13}, \tag{1.132}$$

$$l_{33} = M^{R}_{11}M^{R}_{22}(M_{11} + M_{22}), \tag{1.133}$$

$$l_{34} = -M_{13}M^{R}_{13}\left(M^{R}_{11} + M^{R}_{22}\right), \tag{1.134}$$

$$l_{35} = -M^{R}_{22}\left(M_{11}M^{R}_{13} - 2M^{R}_{11}M_{13} + M_{22}M^{R}_{13}\right), \tag{1.135}$$

$$l_{36} = -M^{R}_{11}\left(M_{22}M^{R}_{13} - 2M^{R}_{22}M_{13} + M_{11}M^{R}_{13}\right), \tag{1.136}$$

$$l_{41} = 0 \tag{1.137}$$

$$l_{42} = M_{22}\left(M^{R}_{13}\right)^2, \tag{1.138}$$

$$l_{43} = -M_{13}M^{R}_{22}M^{R}_{13}, \tag{1.139}$$

$$l_{44} = M_{22}\left[\left(M^{R}_{22}\right)^2 + M^2_{11}M^2_{22} - \left(M^{R}_{13}\right)^2\right], \tag{1.140}$$

$$l_{45} = M_{13}\left[\left(M^{R}_{22}\right)^2 + M^2_{11}M^2_{22} - \left(M^{R}_{13}\right)^2\right], \tag{1.141}$$

$$l_{35} = -M_{13}^{R}\left(M_{13}M_{13}^{R} - 2M_{22}M_{22}\right), \tag{1.142}$$

$$l_{51} = M_{13}\left[\left(M_{22}^{R}\right)^2 + M_{11}^2 M_{22}^2 - \left(M_{13}^{R}\right)^2\right], \tag{1.143}$$

$$l_{52} = M_{13}\left(M_{13}^{R}\right), \tag{1.144}$$

$$l_{53} = -M_{22}M_{13}^{R}\left(M_{11} + M_{22}\right), \tag{1.145}$$

$$l_{54} = M_{13}M_{22}^{R}\left(M_{11}^{R} + M_{22}^{R}\right), \tag{1.146}$$

$$\begin{aligned} l_{55} = {} & M_{22}\left[\left(M_{22}^{R}\right)^2 + M_{11}^2 M_{22}^2 - \left(M_{13}^{R}\right)^2\right] \\ & + M_{22}^{R}\left(M_{11}M_{22}^{R} + M_{11}M_{11}^{R} - 2M_{13}M_{13}^{R}\right) - M_{11}\left(M_{13}^{R}\right)^2, \end{aligned} \tag{1.147}$$

$$l_{56} = M_{13}^{R}\left(M_{11}M_{13}^{R} + M_{22}M_{13}^{R} - 2M_{13}M_{22}^{R}\right), \tag{1.148}$$

$$l_{61} = M_{13}\left(M_{13}^{R}\right)^2, \tag{1.149}$$

$$l_{62} = M_{13}\left[M_{11}^2 M_{22}^2 + \left(M_{11}^{R}\right)^2 - \left(M_{13}^{R}\right)^2\right], \tag{1.150}$$

$$l_{63} = -M_{11}^{R}M_{13}^{R}\left(M_{11} + M_{22}\right), \tag{1.151}$$

$$l_{64} = M_{13}M_{11}^{R}\left(M_{11}^{R} + M_{22}^{R}\right), \tag{1.152}$$

$$l_{65} = M_{13}^{R}\left(M_{11}M_{13}^{R} + M_{22}M_{13}^{R} - 2M_{11}^{R}M_{13}\right), \tag{1.153}$$

$$\begin{aligned} l_{66} = {} & M_{22}\left[M_{11}^{R}M_{22}^{R} + \left(M_{11}^{R}\right)^2 - \left(M_{13}^{R}\right)^2\right] \\ & + M_{11}^{R}\left(M_{11}M_{22}^{R} + M_{11}M_{11}^{R} - 2M_{13}M_{13}^{R}\right) - M_{11}\left(M_{13}^{R}\right)^2. \end{aligned} \tag{1.154}$$

Exercises

1. With the aid of a computer algebra system like MAPLE, derive Equations (1.33) and (1.34) explicitly.
2. Instead of using the hypothesis of elastic energy equivalence, we want to use the hypothesis of strain equivalence, i.e. $\varepsilon_{ij} = \bar{\varepsilon}_{ij}$. In this case, derive the expression for the effective modulus of elasticity $\bar{E}_{ijkl}$ that corresponds to Equation (1.67).
3. Use the following criterion of damage evolution to replace Equation (1.79)

$$g(\bar{\sigma}, B) = \frac{1}{2}J_{ijkl}\bar{\sigma}_{ij}\bar{\sigma}_{kl} - B(\beta) - \beta_0^2$$

Derive the Equation of damage evolution that corresponds to Equation (1.84).

4. Plot the relation between the extensional strain ε_1 and the damage variable ϕ_1 given in Equation (1.102). Make several plots for several values of the constant c_1.
5. Using Equation (1.117a), plot the relation between the damage variable ϕ_1 and the stiffness degradation ratio $\frac{\bar{E}_1}{E_1}$.
6. Using Equation (1.117d), plot the relation between the damage variables ϕ_1 and ϕ_2 with the degradation ratio $\frac{\bar{G}_{12}}{G_{12}}$ of the shear modulus. You should obtain a three-dimensional plot in this case.
7. Derive explicitly the coefficients l_{ij} of Equations (1.122) – (1.154) for the strain concentration factors $\bar{A}^M_{ijkl}$ and $\bar{A}^F_{ijkl}$.

2 Finite Element Damage Analysis of Plate Bending

2.1 Introduction

A ductile material is capable of undergoing large plastic deformations. The accumulated plastic deformation can induce the changes of microstructures of the material through, for example, the nucleation, growth and coalescence of microvoids. These changes in material microstructures are the irreversible thermodynamic processes and result in a progressive degradation on the material properties. The process of the initiation and growth of microvoids and other microdefects induced by plastic deformations in ductile solids is called the ductile plastic damage. The primary interest of the ductile plastic damage is to study the influence of microvoids resulting from plastic deformations on the degradation of material properties. The changes on material properties can be studied by either a phenomenological damage model or a micromechanical damage model. A number of damage definitions and measures were proposed for both the models (vide the review papers of Krajcinovic, 1984, 1989; Chaboche, 1988; among others). Within the framework of phenomenological damage model, the damage of a material can be measured in macroscale by the deduction of mechanical properties, such as the elasticity constants (Lemaitre et al., 1979). Moreover, the changes of the macro-mechanical properties can be characterized by the damage effect parameters which are able to be determined from experiments (Lemaitre, 1985). These damage parameters are the internal state variables in thermodynamics. The phenomenological damage model in conjunction with thermodynamics is not only simple in material modeling, but also quite accurate for the representation of a damage process. Therefore, the phenomenological model is very attractive in the practical application of the damage mechanics for engineering structures. The present study is based on the phenomenological damage model.

Quite a large number of papers on continuum damage mechanics have been published (see the references given in reviews of Krajcinovic, 1984, 1989; Chaboche, 1988). However, the ductile plastic damage of plate bending has received little attention up to now even though some damage models for the bending analysis of brittle beams have been proposed, e.g., Krajcinovic (1979). The flexural plates made of ductile metals, a very important type of structure, may undergo large plastic deformations under certain boundary and loading conditions. The large plastic deformations in a metal plate can induce the initiation and growth of microvoids and consequently cause the deterioration of the mechanical properties of the plate, a damage process. The load-carrying capacity of the damaged plate is lower than the one predicted from the elastoplastic analysis. Therefore, the ductile plastic damage analysis can provide a useful tool for a safe design of metal plates.

In the present computational model, the damage effect parameters of a cross-section are introduced from irreversible thermodynamics to take into account the damage effect across the plate thickness. Analogous to the concept of effective stresses, the effective stress couples are defined for plate bending problems. The yield function is then defined in the effective stress couple space. The evolution law of ductile plastic damage proposed by Lemaitre (1985), in which the damage evolution is a linear function of the equivalent plastic strain, is adopted. The concept of the plastic node model presented by Shi and Voyiadjis (1992) is extended here to discretize the distribution of the damage matrix in an element. Finally, by using the principle of virtual work together with the damage node model proposed here, the elastoplastic-damage stiffness matrix of element is derived. The resulting element stiffness matrix can be obtained explicitly as long as the elastic part of the element stiffness matrix is given explicitly. Consequently, the computational model presented here is very simple and efficient for the damage analysis of elastoplastic bending plates.

The damage-related matrices in the elastoplastic-damage stiffness matrix are dependent on the damage effect matrix (or tensor). The damage model used for the application presented here is the scalar isotropic damage model which is the simplest and most widely used model for the one-dimensional and isotropic phenomenological damage (vide, e.g., the review papers of Chaboche, 1988 and Krajcinovic, 1989). In this model, the change of macromechanical properties of a material caused by microdefects is described by a simple scalar variable: a damage parameter. A new damage strain release rate proposed by the authors (Shi and Voyiadjis, 1993, 1997), in which the influence of damage on the plastic deformations is taken into account, is used in this work. This damage strain release rate can be defined in the effective stress couple space.

The four-noded quadrilateral (12 dof) C^0 strain element for plate bending developed by the authors (Shi and Voyiadjis, 1991) is employed here to evaluate the elastic stiffness matrix. This assumed strain plate element is based upon the shear deformable plate theory proposed by Voyiadjis and Shi (1991a) and the quasi-conforming element method presented by Tang et al.

(1980). Unlike most C^0 plate elements where the element stiffness matrix is evaluated by numerical integration, the element stiffness matrix of the C^0 plate element used here is given explicitly. Consequently, the assumed strain C^0 plate element presented by the authors is very computationally efficient. Furthermore, this four-noded quadrilateral (12 dof) C^0 plate element possesses a linear bending strain field and is free of shearing locking and numerical ill-condition. Therefore, this finite element is capable of giving reliable and accurate results for both thick and thin plate analysis.

2.2 Ductile Plastic Damage in Plate Bending Problems

The damage analysis presented here is based on the phenomenological method. As mentioned earlier, so far there is not much information about the ductile plastic damage analysis of plate bending in the literature. As an earlier step towards the ductile damage analysis of plates, the following assumptions are adopted in the present study for simplicity:

1. The damage process in ductile plastic damage is induced by plastic deformations;
2. Tension and compression have the same influence on the damage development (Lemaitre, 1985);

The first assumption is reasonable for ductile materials since the degradation of elastic modulus in elastic range is really negligible. The second assumption is similar to the plastic behavior of ductile materials. It has some limitations in application. Nevertheless, when the degradation of mechanical properties of a ductile material is primarily induced by microvoids rather than by microcracks, this assumption will be quite feasible.

In order to demonstrate the ductile plastic damage of plate bending more easily, the corresponding one-dimensional case will be examined first.

2.2.1 Ductile Plastic Damage of Beams

Within the framework of the phenomenological damage, the measure of damage can be characterized by the degradation of the elastic modulus of the material (Lemaitre, 1985; Voyiadjis, 1988; Li et al., 1990). If one lets E_0 be the Young's modulus of the material in the undamaged state (virgin material), then the instantaneous Young's modulus E can be determined by the damage effect parameter $d (0 \leq d \leq 1)$ as

$$E = (1 - d)E_0. \tag{2.1}$$

Since the stress induced by bending is non-uniformly distributed along the thickness direction z, the damage parameter d, in general, varies through the thickness of a beam, i.e., $d = d(z)$.

Let $\varepsilon_e(z)$ be the elastic axial strain of the cross-section of a beam, $\sigma(z)$ be the corresponding axial stress. According to the assumption of the plane cross-section, $\varepsilon_e(z)$ can be expressed in terms of the elastic rotation of cross-section ϕ as

$$\varepsilon_e(z) = \frac{\partial \phi}{\partial x} z. \tag{2.2}$$

The corresponding stress then is given by the Hooke's law as follows

$$\sigma(z) = E\varepsilon_e = (1-d)E_0\varepsilon_e = (1-d)E_0 \frac{\partial \phi}{\partial x} z \tag{2.3}$$

where x is the coordinate in the axial direction of the beam. It should be noted that $\sigma(z)$ might be nonlinear across the thickness since $d(z)$ can also be a function of z.

The elastic bending strain energy density of the beam, W_e, is of the form

$$W_e = \frac{b}{2} \int_{-h/2}^{h/2} \varepsilon_e \sigma dz = \frac{b}{2} \frac{\partial \phi}{\partial x} \int_{-h/2}^{h/2} \sigma z dz = \frac{1}{2} \frac{\partial \phi}{\partial x} M \tag{2.4}$$

with

$$M = b \int_{-h/2}^{h/2} \sigma z dz = b \frac{\partial \phi}{\partial x} E_0 \int_{-h/2}^{h/2} [1 - d(z)] z^2 dz \tag{2.5}$$

in which h and b are respectively the thickness and width of the beam. If one lets

$$\bar{\sigma} = E_0 \varepsilon_e = \frac{E}{1-d} \varepsilon_e = \frac{\sigma}{1-d}, \tag{2.6}$$

then the stress couple also takes the form

$$M = b \int_{-h/2}^{h/2} [1 - d(z)] \bar{\sigma} z dz = \bar{M} - b \int_{-h/2}^{h/2} d(z) \bar{\sigma} z dz \tag{2.7a}$$

with

$$\bar{M} = b \int_{-h/2}^{h/2} \bar{\sigma} z dz = bE_0 \frac{\partial \phi}{\partial x} \int_{-h/2}^{h/2} z^2 dz = \frac{\partial \phi}{\partial x} \frac{bh^3}{12}. \tag{2.7b}$$

$\bar{\sigma}$ in Equation (2.6) is the so-called effective stress. The quantity $\bar{M}$ defined in Equation (2.7b) can be considered as the effective stress couple. In the beam bending problem considered here, the product of $\bar{\sigma}$ and z does not change its sign across the beam thickness $-h/2 \leq z \leq h/2$. Therefore, by using the weighted mean-value theorem for integrals, the last term in Equation (2.7a) can be expressed as

$$b \int_{-h/2}^{h/2} d(z)\bar{\sigma} z dz = bd(\xi) \int_{-h/2}^{h/2} \bar{\sigma} z dz = d^* \bar{M} \quad \text{for some } \xi \text{ in } [-h/2, h/2]. \tag{2.8}$$

The parameter d^* in the above Equation is the mean value of damage parameter $d(z)$ across the beam thickness. In the present study, d^* is taken as the damage effect parameter of a cross-section of the beam. The determination of d^* will be presented later. Substituting Equation (2.8) into Equation (2.7a) gives

$$M = (1 - d^*)\bar{M}. \tag{2.9}$$

The rate form of Eq. (2.9) can be written as

$$M = (1 - d^*)\dot{\bar{M}} - \dot{d}^* \bar{M} \tag{2.10}$$

where the symbol "·" signifies the material rate.

Similar to the strain energy release rate in fracture mechanics, which is used for the fracture criterion, a damage strain energy release rate associated with a unit damage growth is defined in damage mechanics (Lemaitre, 1985). From thermodynamics, there is an internal variable, named D here for a one-dimensional problem, corresponding to the damage strain energy release rate $-Y$. By taking the free-energy Ψ as the thermodynamic potential, the damage strain energy release rate $-Y$ of a beam can be defined as

$$Y = \frac{\partial \Psi}{\partial D}. \tag{2.11}$$

In general, the free energy Ψ is the function of elastic strains ε_e, equivalent plastic strain $\bar{\varepsilon}_p$, internal variables $\mathbf{d}$, and absolute temperature T, i.e.,

$$\Psi = \Psi_1(\varepsilon_e, \bar{\varepsilon}_p, \mathbf{d}, T). \tag{2.12}$$

Ψ can also be expressed in terms of effective stresses $\bar{\sigma}$ as

$$\Psi = \Psi_2(\bar{\sigma}, \bar{\varepsilon}_p, \mathbf{d}, T). \tag{2.13a}$$

For the one-dimensional problem considered here, the above Equation takes the form

$$\Psi = \Psi(\bar{M}, \kappa_p, D, T) \tag{2.13b}$$

where κ_p is the plastic curvature. In damage mechanics, the internal variables **d** are the damage effect parameters. It should be noted that both elastic and plastic strains in a damaged material are associated with the damage parameters **d** even in the case where the elastic and plastic free energies can be decoupled.

The damage evolution D can be determined by the normality property of the dissipation potential. By defining a suitable dissipation potential, Lemaitre (1985) proposed a damage evolution model for ductile plastic damage as follows

$$\dot{D} = \left(\frac{-Y}{S_0}\right)^{s_0} \dot{\varepsilon}_p \tag{2.14}$$

with

$$-Y\dot{D} \geq 0 \tag{2.15}$$

$$\dot{D} = 0 \quad \text{when} \quad -Y \leq Y_c \tag{2.16}$$

$$D = D_c \quad \text{ruptured} \tag{2.17}$$

in which S_0 and s_0 are the material constants; Y_c is the critical value of the damage strain energy release rate $-Y$; D_c is the critical value of the damage parameter D. Both Y_c and D_c are material constants and can be determined from experiments. The damage parameter D then can be evaluated from $\dot{D}$ by integration.

When the stresses rather than stress resultants are used, Equation (2.11) takes the form

$$Y = \frac{\partial \Psi}{\partial \mathbf{d}}. \tag{2.18}$$

Because of the analogy between Equations (2.6) and (2.9) as well as between Equations (2.18) and (2.11), it is feasible to assume $d^* = D$. From now on, d^* will be used to represent the damage parameter of a cross-section determined from the free energy in terms of stress couples.

2.2.2 Ductile Plastic Damage of Plate Bending

In the plane stress problem of plate bending, the in-plane strains ε and stresses σ can be written in the matrix forms, respectively, as

$$\varepsilon = \begin{Bmatrix} \varepsilon_x \\ \varepsilon_y \\ 2\varepsilon_{xy} \end{Bmatrix}, \quad \sigma = \begin{Bmatrix} \sigma_x \\ \sigma_y \\ \sigma_{xy} \end{Bmatrix} \tag{2.19a}$$

For a linear elastic, isotropic material, the stresses σ and elastic strains ε_e in the undamaged state have the following relation

$$\sigma = S_0 \varepsilon_e \tag{2.19b}$$

with

$$S_0 = \frac{E_0}{1-\nu_0^2}\begin{bmatrix} 1 & \nu_0 & 0 \\ \nu_0 & 1 & 0 \\ 0 & 0 & \frac{1-\nu_0}{2} \end{bmatrix} \tag{2.20}$$

in which ν_0 is the Poisson's ratio of the undamaged material. After the material is subjected to damage, the corresponding damaged elasticity matrix $S(d)$ can be written as (Krajcinovic, 1989)

$$\mathbf{S}(\mathbf{d}) = [\mathbf{I} - \mathbf{D}(\mathbf{d})]\mathbf{S}_0 \tag{2.21}$$

where $\mathbf{I}$ is the identity matrix and $\mathbf{D}(\mathbf{d})$ signifies the damage effect matrix (or tensor) which is symmetric and composed of damage parameters $\mathbf{d}$. The expression of $\mathbf{D}(\mathbf{d})$ can be determined from suitable micromechanical models (Krajcinovic, 1989). The evolution of the damage parameters $\mathbf{d}$ can be evaluated from the free energy as described earlier.

The stress-strain relation in a damaged state takes the form

$$\sigma = \mathbf{S}(\mathbf{d})\varepsilon_e = [\mathbf{I} - \mathbf{D}(\mathbf{d})]\mathbf{S}_0\varepsilon_e = [\mathbf{I} - \mathbf{D}(\mathbf{d})]\bar{\sigma} \tag{2.22}$$

with the introduction of the effective stresses defined as

$$\bar{\sigma} = \mathbf{S}_0 \varepsilon_e. \tag{2.23}$$

The increment of σ can be expressed as

$$\Delta\sigma = [\mathbf{I} - \mathbf{D}(\mathbf{d})]\Delta\bar{\sigma} - \Delta\mathbf{D}(\mathbf{d}, \Delta\mathbf{d})\bar{\sigma} \tag{2.24}$$

in which $\Delta\mathbf{D}$ is the increment of $\mathbf{D}$ and is composed of $\mathbf{d}$ and $\Delta\mathbf{d}$.

Under the assumption of the plane cross-section, the elastic strains ε_e across a cross-section can be expressed as

$$\varepsilon_e = \begin{Bmatrix} \dfrac{\partial\phi_{xe}}{\partial x} \\ \dfrac{\partial\phi_{ye}}{\partial y} \\ \dfrac{\partial\phi_{xe}}{\partial x} + \dfrac{\partial\phi_{ye}}{\partial y} \end{Bmatrix} z = \kappa_e z \tag{2.25}$$

where ϕ_{xe} and ϕ_{ye} are the generalized elastic rotations of the cross-section at $x =$ constant and $y =$ constant, respectively, and κ_e signifies the elastic curvatures of the plate. The incremental form of Equation (2.25) is as follows

$$\Delta\varepsilon_e = \begin{Bmatrix} \dfrac{\partial\Delta\phi_{xe}}{\partial x} \\ \dfrac{\partial\Delta\phi_{ye}}{\partial y} \\ \dfrac{\partial\Delta\phi_{xe}}{\partial x} + \dfrac{\partial\Delta\phi_{ye}}{\partial y} \end{Bmatrix} z = \Delta\kappa_e z. \tag{2.26}$$

Consequently, the increment of the elastic bending strain energy density of a plate, ΔW_e, can be expressed as

$$\begin{aligned}\Delta W_e &= \frac{1}{2}\int_{-h/2}^{h/2} \Delta\varepsilon_e^T \Delta\sigma dz = \frac{1}{2}\Delta\kappa_e^T \int_{-h/2}^{h/2} [(\mathbf{ID})\Delta\bar{\sigma} - \Delta\mathbf{D}\bar{\sigma}]zdz \\ &= \frac{1}{2}\Delta\kappa_e^T\left[(\mathbf{ID}^*)\Delta\bar{\mathbf{M}} - \Delta\mathbf{D}^*\bar{\mathbf{M}}\right] = \frac{1}{2}\Delta\kappa_e^T\Delta\mathbf{M}\end{aligned} \tag{2.27}$$

in which the right superscript T signifies the matrix transpose and

$$\Delta\bar{\mathbf{M}} = \int_{-h/2}^{h/2} \Delta\bar{\sigma}zdz = \frac{h^3}{12}\mathbf{S}_0\Delta\kappa_e = \mathbf{S}_{b0}\Delta\kappa_e \tag{2.28}$$

$$\mathbf{D}^*\Delta\bar{\mathbf{M}} = \mathbf{D}(\mathbf{d}(\xi))\Delta\bar{\mathbf{M}} = \int_{-h/2}^{h/2} \mathbf{D}(\mathbf{d}(z))\Delta\bar{\sigma}zdz, \tag{2.29}$$

some ξ in $[-h/2, h/2]$

$$\Delta\mathbf{D}^*\bar{\mathbf{M}} = \Delta\mathbf{D}^*(\Delta\mathbf{d}(\xi), \mathbf{d}(\xi))\bar{\mathbf{M}} = \int_{-h/2}^{h/2} \Delta\mathbf{D}(\mathbf{d}(z), \Delta\mathbf{d}(z))\bar{\sigma}zdz \tag{2.30}$$

In Equations (2.29) and (2.30), the weighted mean-value theorem for integrals is utilized. The parameters $\mathbf{d}(\xi)$ and $\Delta\mathbf{d}(\xi)$ are the mean values of damage parameter $\mathbf{d}(z)$ and its increment $\Delta\mathbf{d}(z)$ along the plate thickness respectively. $\mathbf{d}(\xi)$ and $\Delta\mathbf{d}(\xi)$ are defined as, respectively, the damage effect parameter and its increment of a plate at the point of the plate under consideration. Similar to the one-dimensional problem, $\Delta\mathbf{d}(\xi)$ and $\mathbf{d}(\xi)$ can also be evaluated by the internal variables $\mathbf{D}_I$ corresponding to the damage strain energy release rates $-\mathbf{Y}$. For plate bending problem, $-\mathbf{Y}$ takes the form $-\mathbf{Y} = \{Y_1, Y_2\}$, and is given by the free energy of the plate $\Psi = \Psi(\bar{\mathbf{M}}, \bar{\kappa}_p, \mathbf{D}_I, T)$ as

$$\mathbf{Y} = \frac{\partial\Psi}{\partial\mathbf{D_I}}. \tag{2.31}$$

The increments of the damage effect parameters of a cross-section, $\Delta\mathbf{d}(\xi)$, are given by

$$\Delta\mathbf{d}^* = \Delta\mathbf{d}(\xi) = \Delta\mathbf{D}_I = \left(\frac{1}{S_0}\right)^{S_0}\begin{Bmatrix}-Y_1^{S_0}\\ -Y_2^{S_0}\end{Bmatrix}\Delta\bar{\kappa}_p \tag{2.32}$$

where $\Delta\bar{\kappa}_p$ is the increment of the equivalent plastic curvature. Let $\Delta\bar{\kappa}_{xp}$, $\Delta\bar{\kappa}_{yp}$, and $\Delta\bar{\kappa}_{xyp}$, be the increments of plastic curvatures, then $\Delta\bar{\kappa}_p$ can be expressed as

$$\Delta\bar{\kappa}_p = \frac{2}{3^{1/2}}\left(\Delta\kappa_{xp}^2 + \Delta\kappa_{yp}^2 + \Delta\kappa_{xp}\Delta\kappa_{yp} + \Delta\kappa_{xyp}^2\Big/4\right)^{1/2}. \tag{2.33}$$

Following the concept used in the three-dimensional problem presented by Lee et al. (1985), the damage criterion of a plate can be written as

$$F_d(\mathbf{Y}, \beta) = \frac{1}{2}\left(\mathbf{Y}^T\mathbf{J}\mathbf{Y}\right)^{1/2} - [B_0 + B(\beta)] = 0 \tag{2.34}$$

in which $\mathbf{J}$ is a symmetric matrix; B_0 is the initial damage threshold; $B(\beta)$ is the damage threshold strengthening; and β is an overall damage parameter. The determination of $\mathbf{J}$ can be found in Lee et al. (1985) and Chaw and Lu (1989). B_0 and $B(\beta)$ can be obtained from experiments (Chaw and Wang, 1988). The increment of β can be expressed as (Lee et al., 1985)

$$\Delta\beta = 2\left(\frac{1}{2}\Delta\mathbf{D}_I^T\mathbf{J}\Delta\mathbf{D}_I\right)^{1/2}. \tag{2.35}$$

The following expression can be used as a rupture criterion (Lee et al, 1985)

$$\beta = \Sigma\Delta\beta = \beta_c \tag{2.36}$$

where β_c is the critical value of overall damage which is a material constant and can be obtained from experiments (Chaw and Wang, 1988).

2.3 A Damage Node Model for Ductile Plastic Damage of Plate Bending

A new plastic node model for the finite element plastic analysis of plates and shells was presented (Shi and Voyiadjis, 1992a). In this plastic node model, the yield function, in terms of stress couples and stress resultants, is checked only at the element nodes. When the stress couples and stress resultants at a node satisfy the yield function, the node of the element is considered to become a plastic node. The plastic deformations are developed only at these plastic nodes, and the interior of the element is always in elastic. The concept of plastic nodes is extended to damage analysis in the present work. That is, the damage criterion is only checked at the element nodes and it is assumed that the damage only undergoes at the damaged nodes.

Because of the nonlinear nature of the plastic analysis, the incremental scheme is used in the evaluation of the elasto-plastic stiffness matrix. It is assumed in this work that the incremental bending strains of a plate, i.e., incremental curvature $\Delta\kappa$ can be decomposed into two parts: elastic parts $\Delta\kappa_e$ and plastic parts $\Delta\kappa_p$, i.e.,

$$\Delta\kappa = \Delta\kappa_e + \Delta\kappa_p. \tag{2.37}$$

Similarly, the incremental nodal displacement vector of an element $\Delta\mathbf{q}$ takes the form

$$\Delta\mathbf{q} = \Delta\mathbf{q}_e + \Delta\mathbf{q}_p. \tag{2.38}$$

In the finite element modeling of plate bending using the generalized displacement method, $\Delta\kappa_e$ in an element can be expressed in terms of the strain-displacement matrix **B** and the nodal elastic displacement vector of the element $\Delta\mathbf{q}_e$ as

$$\Delta\kappa_e(x,y) = \mathbf{B}(x,y)\Delta\mathbf{q}_e = \mathbf{B}\Delta\mathbf{q}_e. \tag{2.39}$$

If one lets S_b be the flexural rigidity matrix, the element stress couples $\Delta\mathbf{M}$ are then given by

$$\Delta\mathbf{M}(x,y) = \begin{Bmatrix} \Delta M_x \\ \Delta M_y \\ \Delta M_{xy} \end{Bmatrix} = \mathbf{S}_b\Delta\kappa_e(x,y) = \mathbf{S}_b\mathbf{B}\Delta\mathbf{q}_e. \tag{2.40}$$

Consequently, the elastic stiffness matrix of an element, $\mathbf{K}_e$, can be written as

$$\mathbf{K}_e = \int_\Omega \mathbf{B}^T\mathbf{S}_b\mathbf{B}dxdy \tag{2.41}$$

where Ω denotes the element domain.

The plastic curvatures can be obtained from the yield function and the associated flow rule. If one lets $F_i(\mathbf{M}_i)$ be the yield function, and $d\lambda_i$ be the plastic proportionality parameter at node i, then by recalling that the plastic deformations are only developed at the plastic nodes, the increment of the plastic curvature κ_{xp} in an element is given by

$$\Delta\kappa_{xp}(x,y) = \sum_{i=1}^{NPN} \delta(x - x_i, y - y_i)d\lambda_i \frac{\partial F_i}{\partial M_{xi}} \tag{2.42}$$

in which *NPN* denotes the number of plastic nodes in the element under consideration.

By using the variational principle to determine $d\lambda_i$ in terms of elastic nodal displacement vector $\Delta\mathbf{q}$ (Shi and Voyiadjis, 1992a), then the elasto-plastic stiffness matrix of a finite element, $\mathbf{K}_{ep}$, takes the form

$$\mathbf{K}_{ep} = \mathbf{K}_e\left[\mathbf{I} - \mathbf{a}\left(\mathbf{a}^T\mathbf{K}_e\mathbf{a} + \mathbf{H}\right)^{-1}\mathbf{a}^T\mathbf{K}_e\right] \tag{2.43}$$

where **a** is the plastic nodal displacement matrix which is associated with the yield function in terms of the stress couples, and **H** is a matrix related to the plastic stiffness of the given material (vide Shi and Voyiadjis, 1992a).

It is worthwhile to mention that $\mathbf{K}_{ep}$ presented here can be evaluated explicitly, i.e., without numerical integration, when the elastic stiffness matrix $\mathbf{K}_e$ can be given explicitly. This feature makes the present plastic node model very computationally efficient and attractive.

When a material is subjected to a damage process, the mechanical properties of the material are degraded. Consequently, the yield stress of the damaged material decreases as the damage increases. However, when the stress σ are replaced by the effective stresses $\bar{\sigma}$ which are associated with the virgin material, the yield function of the virgin material can be used for the damaged material. For example, in the plastic damage analysis of plates, the yield function takes the form

$$F\left(\mathbf{M}, \sigma_{yd}(\mathbf{D}), k\right) = \bar{F}\left(\bar{\mathbf{M}}, \sigma_{yo}, k_0\right) = 0 \tag{2.44}$$

where $\mathbf{M}$ is the stress couple vector of the plate; $\sigma_{yd}(\mathbf{D})$ is the yield stress of the damaged material which is the function of the damage tensor $\mathbf{D}$; $\bar{\mathbf{M}}$ is the effective stress couple; σ_{yo} is the yield stress of the virgin material; and k and k_0 are the strain hardening parameters of the damaged and virgin materials, respectively.

From the associated flow rule, the incremental plastic curvatures at node i are given by

$$\Delta\kappa_{p_i} = \frac{\partial \bar{F}_i}{\partial \mathbf{M}_i} d\lambda_i = \frac{\partial \bar{\mathbf{M}}_i}{\partial \mathbf{M}_i}\frac{\partial \bar{F}_i}{\partial \bar{\mathbf{M}}_i} d\lambda_i = \left(\mathbf{ID}_i^*\right)^{-1}\frac{\partial \bar{F}_i}{\partial \bar{\mathbf{M}}_i} d\lambda_i \quad \text{(no summation on } i) \tag{2.45}$$

in which Equations (2.22) and (2.27) are used. It should be noted that there is no summation on the repeated indices in this work.

For an element in the damaged state, corresponding to a virtual nodal displacement vector $\delta\Delta\mathbf{q} = \delta\Delta\mathbf{q}_e + \delta\Delta\mathbf{q}_p$ and a virtual bending strain field $\delta\Delta\kappa = \delta\Delta\kappa_e + \delta\Delta\kappa_p$, the principle of virtual work for the element gives

$$\begin{aligned} \delta\Delta\mathbf{q}^T\Delta\mathbf{f} &= \int_\Omega \delta\Delta\kappa^T\Delta\mathbf{M}dxdy = \int_\Omega \left(\delta\Delta\kappa_e^T + \delta\Delta\kappa_p^T\right)\left[(\mathbf{ID}^*)\Delta\bar{\mathbf{M}} - \Delta\mathbf{D}^*\bar{\mathbf{M}}\right]dxdy \\ &= \int_\Omega \Big[\delta\Delta\kappa_e^T\bar{\mathbf{M}} + \delta\Delta\kappa_p^T(\mathbf{ID}^*)\Delta\bar{\mathbf{M}} - \delta\Delta\kappa_e^T\mathbf{D}^*\Delta\bar{\mathbf{M}} - \\ &\qquad \delta\Delta\kappa_e^T\Delta\mathbf{D}^*\bar{\mathbf{M}} - \delta\Delta\kappa_p^T\Delta\mathbf{D}^*\bar{\mathbf{M}}\Big]\, dxdy \end{aligned} \tag{2.46}$$

where $\Delta\mathbf{f}$ represents the increment of the internal nodal force vector of the element. In Equation (2.46), $\mathbf{D}^*(x, y)$ and $\Delta\mathbf{D}^*(x, y)$ are yet undefined fields.

Similar to the concept of the plastic nodes, a damage node model is proposed here to construct $\mathbf{D}^*(x, y)$ and $\Delta\mathbf{D}^*(x, y)$. In this damage node model, the damage is assumed to be developed at the element nodes only, and the interior of the element is always in the undamaged state. Consequently, the damage matrix or tensor in an element $\mathbf{D}^*$ and $\Delta\mathbf{D}^*$ can be expressed as

$$\mathbf{D}^*(x, y) = \sum_{j=1}^{NDN} \delta\left(x - x_j, y - y_j\right)\mathbf{D}_j^* \tag{2.47}$$

$$\Delta\mathbf{D}^*(x, y) = \sum_{j=1}^{NDN} \delta\left(x - x_j, y - y_j\right)\Delta\mathbf{D}_j^* \tag{2.48}$$

where $\mathbf{D}_j^*$ and $\Delta\mathbf{D}_j^*$ are the damage matrix and its increment at node j of the element under consideration; and NDN represents the number of damage nodes of the element. Substituting Equations (2.41), (2.45), (2.47) and (2.48) into Equation (2.46) leads to

$$\begin{aligned}\delta\Delta\mathbf{q}^T\Delta\mathbf{f} &= \delta\Delta\mathbf{q}_e^T\mathbf{K}_e\Delta\mathbf{q}_e + \delta d\lambda^T\mathbf{H}d\lambda \\ &\quad - \sum_{j=1}^{NDN}\left[\delta\Delta\kappa_{ej}^T\mathbf{D}^*\Delta\bar{\mathbf{M}}_j + \delta\Delta\kappa_{ej}^T\Delta\mathbf{D}_j^*\bar{\mathbf{M}}_j + \delta d\lambda_j\frac{\partial\bar{F}_j^T}{\partial\bar{\mathbf{M}}_j}\left(\mathbf{ID}_j^*\right)^{-T}\Delta\mathbf{D}_j^*\bar{\mathbf{M}}_j\right].\end{aligned} \tag{2.49a}$$

In the above derivation, the consistency condition of yield function

$$\frac{\partial\bar{F}_i^T}{\partial\bar{\mathbf{M}}_i}d\bar{\mathbf{M}}_i = -\frac{\partial\bar{F}_i}{\partial k_{0i}}dk_{0i} = H_i d\lambda_i \tag{2.49b}$$

is also utilized. It follows from Equations (2.39) and (2.40) that the elastic bending strains and stress couples at node j of the element can be written as

$$\Delta\kappa_{ej} = \Delta\kappa_e\left(x_j, y_j\right) = \mathbf{B}\left(x_j, y_j\right)\Delta\mathbf{q}_e = \mathbf{B}_{nj}\Delta\mathbf{q}_e \tag{2.50}$$

$$\Delta\bar{\mathbf{M}}_j = \mathbf{S}_{b0}\Delta\kappa_e\left(x_j, y_j\right) = \mathbf{S}_{b0}\mathbf{B}_{nj}\Delta\mathbf{q}_e. \tag{2.51}$$

After some mathematical manipulations, one can obtain

$$\sum_{j=1}^{NDN}\delta\Delta\kappa_{ej}^T\mathbf{D}^*\Delta\bar{\mathbf{M}}_j = \delta\Delta\mathbf{q}_e^T\mathbf{B}_n^T\mathbf{D}_e^*\mathbf{S}_{be}\mathbf{B}_n\Delta\mathbf{q}_e \tag{2.52}$$

in which $\mathbf{B}_n$ is the matrix consisting of $\mathbf{B}_{nj}$ $(j = 1, 2, NDN)$, $\mathbf{D}_e^*$ is the damage matrix of the element, and $\mathbf{S}_{be}$ is the enlarged diagonal matrix of $\mathbf{S}_{b0}$. $\mathbf{D}_e^*$ takes the form

$$\mathbf{D}_e^* = \begin{bmatrix} \mathbf{D}_1^* & 0 & 0 \\ 0 & \mathbf{D}_j^* & 0 \\ 0 & 0 & \mathbf{D}_{NDN}^* \end{bmatrix}. \tag{2.53}$$

In the ductile plastic damage, the increment of the damage tensor $\Delta\mathbf{D}_j^*$ at node j is the function of the damage parameters $\Delta\mathbf{d}_j^*$ at the node, and $\Delta\mathbf{d}_j^*$ depends on the increment of the equivalent plastic curvature $\Delta\kappa_{pj}$ shown in Equation (2.32). By using Equations (2.33) and (2.45), $\Delta\mathbf{D}_j^*\left(\Delta\mathbf{d}_j^*\right)$ can be transformed into

$$\Delta\mathbf{D}_j^*\left(\Delta\mathbf{d}_j^*\right) = \mathbf{D}_1\left(\Delta\kappa_{pj}\right) = \mathbf{D}_2\left(d\lambda_j\right). \tag{2.54}$$

Therefore, in the case when $\Delta\mathbf{D}_j^*$ is a linear function of $\Delta\mathbf{d}_j^*$ after a simple rearrangement the last two terms in Equations (2.49a) can be written as

$$\sum_{j=1}^{NDN} \delta\Delta\kappa_{ej}^T \Delta\mathbf{D}_j^* \bar{\mathbf{M}}_j = \sum_{j=1}^{NDN} \delta\Delta\kappa_{ej}^T \mathbf{D}_2\left(d\lambda_j\right)\bar{\mathbf{M}}_j = \delta\Delta\mathbf{q}_e^T \mathbf{B}_n^T (\mathbf{DM}) d\lambda_j \tag{2.55}$$

$$\sum_{j=1}^{NDN} \delta d\lambda_j \frac{\partial \bar{F}_j^T}{\partial \bar{\mathbf{M}}_j}\left(\mathbf{ID}_j^*\right)^{-T} \Delta\mathbf{D}_j^* \bar{\mathbf{M}}_j = \delta d\lambda^T (\mathbf{DF})^T (\mathbf{ID})(\mathbf{DM}) d\lambda \tag{2.56}$$

with

$$d\lambda^T = \left\{d\lambda_1, d\lambda_j, d\lambda_{NDN}\right\}. \tag{2.57}$$

By substituting Equations (2.52), (2.55), and (2.56), Equations (2.49a) becomes

$$\begin{aligned}
\delta\Delta\mathbf{q}^T\Delta\mathbf{f} &= \delta\Delta\mathbf{q}_e^T \mathbf{K}_e \Delta\mathbf{q}_e + \delta d\lambda^T \mathbf{H} d\lambda - \delta\Delta\mathbf{q}_e^T \mathbf{B}_n^T \mathbf{D}_e^* \mathbf{S}_{be} \mathbf{B}_n \Delta\mathbf{q}_e \\
&\quad - \delta\Delta\mathbf{q}_e^T \mathbf{B}_n^T (\mathbf{DM}) d\lambda - \delta d\lambda^T (\mathbf{DF})^T (\mathbf{ID})(\mathbf{DM}) d\lambda \\
&= \delta\Delta\mathbf{q}_e^T \left[\left(\mathbf{K}_e - \mathbf{B}_n^T \mathbf{D}_e^* \mathbf{S}_{be} \mathbf{B}_n\right)\Delta\mathbf{q}_e - \mathbf{B}_n^T (\mathbf{DM}) d\lambda\right] \\
&\quad + \delta d\lambda^T \left[\mathbf{H} d\lambda - (\mathbf{DF})^T (\mathbf{ID})(\mathbf{DM}) d\lambda\right].
\end{aligned} \tag{2.58}$$

By recalling Equation (2.38), the above Equation can be rewritten as

$$\begin{aligned}
&\delta\Delta\mathbf{q}^T\left[\mathbf{K}_e^*(\Delta\mathbf{q} - \bar{\mathbf{a}} d\lambda)\mathbf{B}_n^T (\mathbf{DM}) d\lambda - \Delta\mathbf{f}\right] \\
&+ \delta d\lambda^T\left[\mathbf{H} d\lambda - (\mathbf{DF})^T (\mathbf{ID})(\mathbf{DM}) d\lambda - \bar{\mathbf{a}}^T \mathbf{K}_e^*(\Delta\mathbf{q} - \bar{\mathbf{a}} d\lambda) + \bar{\mathbf{a}}^T \mathbf{B}_n^T (\mathbf{DM}) d\lambda\right] = 0
\end{aligned} \tag{2.59}$$

where $\bar{\mathbf{a}}$ is the modified plastic nodal displacement matrix because of the damage and $\mathbf{K}_e^*$ is the modified elastic stiffness matrix. Matrix $\bar{\mathbf{a}}$ gives the plastic nodal displacement vector as (Shi and Voyiadjis, 1992a)

$$\Delta\mathbf{q}_p = \bar{\mathbf{a}} d\lambda \tag{2.60}$$

$\mathbf{K}_e^*$ is of the form

$$\mathbf{K}_e^* = \mathbf{K}_e - \mathbf{B}_n^T \mathbf{D}_e^* \mathbf{S}_{be} \mathbf{B}_n. \tag{2.61}$$

Since $\delta\Delta\mathbf{q}$ and $\delta d\lambda$ are independent of each other and arbitrary, Equation (2.59) then gives the following two Equations

$$\mathbf{K}_e^*(\Delta\mathbf{q} - \bar{\mathbf{a}}d\lambda)\mathbf{B}_n^T(\mathbf{DM})d\lambda = \Delta\mathbf{f} \tag{2.62}$$

$$\mathbf{A}d\lambda - \bar{\mathbf{a}}^T\mathbf{K}_e^*\Delta\mathbf{q} = 0 \tag{2.63}$$

where $\mathbf{A}$ is the nonsingular square matrix and of the form

$$\mathbf{A} = \mathbf{H} + \bar{\mathbf{a}}^T\mathbf{K}_e^*\bar{\mathbf{a}} + \bar{\mathbf{a}}^T\mathbf{B}_n^T(\mathbf{DM}) - (\mathbf{DF})^T(\mathbf{ID})(\mathbf{DM}). \tag{2.64}$$

Equation (2.63) gives

$$d\lambda - \mathbf{A}^{-1}\bar{\mathbf{a}}^T\mathbf{K}_e^*\Delta\mathbf{q}. \tag{2.65}$$

Substituting Equation (2.65) into Equation (2.62), one finally obtains

$$\mathbf{K}_{epd}\Delta\mathbf{q} = \Delta\mathbf{f} \tag{2.66}$$

in which $\mathbf{K}_{epd}$ is the elastoplastic-damage stiffness matrix of an element and takes the form

$$\mathbf{K}_{epd} = \mathbf{K}_e^*(\mathbf{I} - \bar{\mathbf{a}}^T\mathbf{A}^{-1}\bar{\mathbf{a}}^T\mathbf{K}_e^*)\mathbf{B}_n^T(\mathbf{DM})\mathbf{A}^{-1}\bar{\mathbf{a}}^T\mathbf{K}_e^*. \tag{2.67}$$

It should be noted that $\mathbf{K}_{epd}$ is, in general, unsymmetric in the presence of damage. It can be seen that $\mathbf{K}_{epd}$reduces to $\mathbf{K}_{ep}$ when the damage tensor is null which results in $\mathbf{D}_e^* = 0$, $(\mathbf{DM}) = 0$ and $(\mathbf{ID}) = \mathbf{I}$. Similar to $\mathbf{K}_{ep}$, $\mathbf{K}_{epd}$ can also be evaluated explicitly as long as the elastic stiffness matrix $\mathbf{K}_e$ is given explicitly. A four-noded strain element with the explicit stiffness matrix for the elastic plate bending analysis can be found in the paper of Shi and Voyiadjis (1991b).

Exercises

1. Consider the following expression for the quantity Q used in elasto-plastic damage analysis - see page 120 of the book by Voyiadjis and Kattan (1999).

$$Q = \frac{\partial f}{\partial \tau_{ab}} E_{abcd} \frac{\partial f}{\partial \tau_{cd}} - \frac{\partial f}{\partial \kappa}\sqrt{\frac{\partial f}{\partial \sigma_{pq}}\frac{\partial f}{\partial \sigma_{pq}}}$$
$$- b\frac{\partial f}{\partial \alpha_{ef}}(\tau_{ef} - \alpha_{ef})\frac{\frac{\partial f}{\partial \sigma_{ij}}\frac{\partial f}{\partial \sigma_{ij}}}{(\tau_{gh} - \alpha_{gh})\frac{\partial f}{\partial \sigma_{gh}}}$$

Derive an appropriate expression for the effective quantity $\bar{Q}$ using the following transformation relations for the stress, deviatoric stress, and backstress tensors:

$$\bar{\sigma}_{ij} = M_{ijkl}\sigma_{kl}$$
$$\bar{\tau}_{ij} = N_{ijkl}\sigma_{kl}$$
$$\bar{\alpha}_{ij} = N_{ijkl}\beta_{kl}$$

Use also the transformation relation for the elastic modulus E_{ijkl} given in Equation (1.67). Derive also an expression for N_{ijkl} in terms of M_{ijkl}.

2. For a composite material consisting of a matrix and fibers under uniaxial tension, derive the following expression for the effective modulus of elasticity – see page 45 of the book by Voyiadjis and Kattan (1999).

$$\bar{E} = \frac{c^M(1-\phi_1^M)\bar{E}^M\bar{C}_1^M + c^F(1-\phi_1^F)\bar{E}^F\bar{C}_1^F}{1-\phi_1}$$

where

c^M, c^F	are matrix and fiber volume fractions
$\bar{E}^M$, $\bar{E}^F$	are effective matrix and fiber elastic modulii
$\bar{C}_1^M$, $\bar{C}_1^F$	are effective matrix and fiber strain concentration factors
ϕ_1^M, ϕ_1^F	are matrix and fiber damage variables in one dimension
ϕ_1	is the overall damage variable in one dimension.

3. For an elasto-plastic material with kinematic hardening, derive an expression for the effective backstress tensor rate $d\bar{\beta}_{ij}$ (see Section 7.4 of the book by Voyiadjis and Kattan, 1999).

3 Using DNA

In this chapter we cover the basic steps needed to install and run the program DNA correctly and obtain the results. A complete list of DNA commands and supported finite elements are given in Chapter 4. A detailed outline of the tensor library is given in Chapter 5.

DNA stands for Damage Nonlinear Analysis. It was developed by the research group of Professor George Z. Voyiadjis at Louisiana State University in Baton Rouge, LA, USA. It includes both the elastic and plastic analysis of materials incorporating damage effects based on the theory of continuum damage mechanics. Both linear and nonlinear analysis options are available in DNA. It was written in Fortran 77 with some Fortran 90 extensions. It is a 32-bit DOS executable file which can run under the Windows 95/98/NT operating system. Two versions of the program are available: DNA_S for problems with small strains, and DNA_L for problems with large (finite) plastic strains. The program includes a well developed library of finite elements for two-dimensional plane stress, plane strain, and axisymmetric problems. In addition, three-dimensional analysis can be carried out using the available three-dimensional finite elements. Also, a DNA Windows Interface is also provided for those users who wish to run DNA from within the Windows environment.

For each problem solved with DNA, two text (ASCII) files are associated. The first one is the input file which is prepared by the user using any text editor. After running the program successfully, a second file (output file) which contains the results is generated. Furthermore, if the graphics option is enabled in the input file, a third file (postscript file) is generated which the user can use to view the finite element mesh and various contour plots graphically. For this last step, a postscript viewer must be available – such a program called Ghostscript is provided on the accompanying CD-ROM. With the aid of this viewer, the results can be viewed graphically on the computer screen and printed on the printer. Complete details about the use of DNA and the three file formats are included in this chapter.

The CD-ROM accompanying this guide contains all the following:

1. Fortran source code files for DNA_S.
2. Fortran source code files for DNA_L.

3. DOS executable file DNA_S.
4. DOS executable file DNA_L.
5. Fortran source code for the tensor library.
6. Ghostscript 5.50 and GSView 2.70 for Windows 95/98.
7. Various input example files.
8. Various output and postscript example files.

The only known limitations of the program are:
1. The number of nodes in a problem must not exceed 3000.
2. The number of elements in a problem must not exceed 400.
3. The number of degrees of freedom at each node is limited to a maximum of 3.
4. The maximum number of materials used in a problem is 10.
5. The number of nodes for each element should not exceed 20.
6. The number of skew (inclined) boundary conditions (supports) should not exceed 300.
7. The number of integration points in any direction for each element has a maximum value of 3.
8. The maximum number of interfacial nodes is 500.
9. The system resources of the user's computer (especially the available memory) may limit the size of the problem to be solved.

The program DNA is designed to run in DOS mode under Windows 95/98/NT or Windows 2000. It will not run under Linux or any other Unix system. However, the source code can be compiled and binaries built to run under any operating system with the appropriate tools. This can be done with only slight modifications to the source code. It is noted that the source code was written in such a way so that it will compile easily using the Digital or Compaq Visual Fortran compiler (version 5 or 6 or later). More details about this issue can be found at the end of this chapter.

DNA is a finite element program for the analysis of damage in two-dimensional and three-dimensional problems of solid mechanics. DNA requires an input file and writes the results to an output file. Both the input and output files are text (ASCII) files that can be viewed and edited with any suitable text editor (e.g. DOS Edit, Windows NotePad, or Windows WordPad). The input file is usually prepared by the user by including the appropriate DNA commands – Chapter 4 provides a comprehensive reference guide to these commands. Once the program is exectued, DNA generates the output file. Optionally, DNA may generate an output graphics file written with the PostScript language, depending on whether the graphics option is enabled or not in the input file.

Table 3.1. Recommended Filename Extensions

Type of File	File Extension	Unit Number
Input File	*.in	Unit 11
Output File (Numerical Data)	*.out	Unit 13
Output File (Graphical Data) – Postscript	*.ps	Unit 4

There are no specific filename formats required for the three files associated with DNA. However, in order to organize things, it is advised that the following filename extensions be used as shown in Table 3.1.

There are two versions of the program: DNA_S for problems with small strains, and DNA_L for problems with finite plastic strains. In the rest of this chapter, DNA_S will be used for the example. However, the user can use DNA_L instead of DNA_S (by substituting DNA_L for each occurrence of DNA_S). It should be noticed that DNA_S runs much faster than DNA_L. In particular, large problems used with DNA_L take a very long time to run depending on the type and power of the computer used (i.e CPU speed and memory available).

3.1 Installing DNA

Although DNA can be executed and run from the accompanying CD-ROM directly, it is preferred to install the program on the hard disk. The advantages of hard disk installation are the increase in access speed and the ample disk space available to store the large output files. The user should note that DNA's disk storage requirements are minimal – it needs less than 4 MB for installation. However, the user should keep ample space available for the output files especially for large problems.

Copy the following files from the CD-ROM to a folder on the hard disk:

DNA_S.EXE	(Small Strains Program)
DNA_L.EXE	(Finite Plastic Strains Program)
SAMPLE.IN	(Sample Input File)

Since both versions of DNA are 32-bit DOS applications, it is more appropriate to execute the program from the DOS prompt. Later in the chapter, we will discuss how to run DNA using the DNA Windows Interface directly from within Windows (see Sections 3.5 and 3.6).

3.2 Running DNA

Change to the DOS prompt on your Windows computer and to the folder where you installed the DNA program files. You can now view or edit the example input file (sample.in) using your favorite text editor. You need to check the input file at this stage. Table 3.2 includes a listing of the sample input file (sample.in) that you should have ready to run.

Note that the commands can be entered in capital or lowercase letters; DNA is not case-sensitive.

At the DOS prompt, type the following command to start the execution of DNA:

DNA_S followed by pressing "Enter"

The computer will ask you now to enter some additional information as follows:

(D)elete or (K)eep intermediate files after successful run [D]>

Usually the intermediate files generated during a DNA session are not needed by the user; therefore, type the following command to confirm deletion of these files:

D followed by pressing "Enter"

If you want to keep the intermediate files, type *K* instead of *D* above. Next the computer requests the name of the input file as follows:

File name missing or blank – please enter file name
UNIT 11?

The program refers to the input file internally as unit 11. Enter the name of the sample input file as follows:

SAMPLE.IN followed by pressing "Enter"

The computer prompts now for the name of the output file as follows:

File name missing or blank – please enter file name
UNIT 13?

Enter the name of the desired output file (unit 13) for numerical data as follows:

SAMPLE.OUT followed by pressing "Enter"

Finally the computer prompts for the name of the output postscript file (unit 4) for graphical data as follows:

File name missing or blank – please enter file name
UNIT 4?

Table 3.2. Sample Input File (sample.in) for the Example Problem

```
TITLE 1
 plane strain problem -- plasticity
/*linear
 nonlinear
 nonsymmetric
outputs every 10
NIPXI 3 NIPETA 3
GRAPHICS 10
 WL 0.0 WR 8.0 WB 0.0 WT 10.0
 FMAG 1.0 DMAG 1.0
 CONTOURS
 LINE THICKNESS 1
END
MATERIAL 1 E 73087.0 NU 0.3 THICK 3.175
TYPE 2
yield 226.85 isotropic 792.92 kinematic 275.8
increments 100 iterations 50 convergence factor 0.0001 stop 50
 NODES 9
  1   0.0   0.0  0
  2   4.0   0.0  0
  3   8.0   0.0  0
  4   0.0   5.0  0
  5   4.0   5.0  0
  6   8.0   5.0  0
  7   0.0   10.0  0
  8   4.0   10.0  0
  9   8.0   10.0  0
BOUNDARY CONDITIONS 9
  1   1 1   0
  2   0 0   0
  3   0 0   0
  4   0 0   0
  5   0 0   0
  6   0 0   0
  7   1 1   0
  8   0 0   0
  9   0 0   0
INCIDENCES 4
1,2204,4, 1 2 5 4,1,0
2,2204,4, 2 3 6 5,1,0
3,2204,4, 4 5 8 7,1,0
4,2204,4, 5 6 9 8,1,0
LOADS 1
  9  10000.0  10000.0  0.0
```

Enter the name of the desired postscript file for graphical data as follows:
SAMPLE.PS followed by pressing "Enter"

The computer will run the problem now and generate the two output files "sample.out" and "sample.ps". In case the GRAPHICS block is commented in the input file (using the operator /*), then there will be no postscript output generated (you will not be prompted for unit 4).

Alternatively, the program can be executed with all commands on one line including all the file names by typing the following at the DOS prompt;
DNA_S D SAMPLE.IN SAMPLE.OUT SAMPLE.PS
followed by pressing "Enter"

At the end of the run, the program will show the CPU usage on the screen (it took 26.91 seconds to run this sample problem on a 450-MHz Pentium III computer with 128 MB RAM).

The output file containing the numerical results (sample.out – on the accompanying CD-ROM) can now be viewed with your favorite text editor. The results shown in the output file include the following information:

1. The strain components for each element.
2. The stress components for each element.
3. The reactions at the nodes.
4. The displacements at the nodes.

3.3 Installing Ghostscript and GSView

In order to be able to view the finite element mesh and contour plots, we need to install the postscript viewer, namely Ghostscript and GSView. Copy the compressed file GSV27550.EXE to a folder on the hard disk. Run this program to unzip the files. Then run the setup (installation) program. Two programs Ghostscript 5.50 and GSView 2.70 should be installed automatically on your computer in the default directory C:\gstools. Now you are ready to run GSView which is the postscript viewer for Windows 95/98.

3.4 Viewing the Finite Element Mesh and Contours

Run GSView from the Programs menu in Windows 95/98, then open the file "SAMPLE.PS". You should be able to view the finite element mesh. If the CONTOURS option is enabled in the input file, you should also be able to view the various contours for this problem. The mesh should be exactly like the one shown in Figure 3.1. The PS file can be transformed into a PDF file using Adobe Distiller which comes with Adobe Acrobat from Adobe.

The graphics file for this sample problem consists of 111 pages where the first page contains the finite element mesh, followed by 11 pages for each of the 10 increment outputs as specified in the input file ($11 \times 10 + 1 = 111$ pages). The 11 pages for each output include the following information:

1. Displaced shape.
2. Stress contours for σ_x.
3. Stress contours for σ_y.
4. Stress contours for τ_{xy}.
5. Stress contours for σ_z.
6. Strain contours for ε_x.
7. Strain contours for ε_y.
8. Strain contours for γ_{xy}.
9. Strain contours for ε_z.
10. Strain contours for ε_v – volumetric strain.
11. Plastic work κ.

The 11 graphical output pages after increment 10 are shown in Figures 3.1–3.11.

Fig. 3.1. Finite Element Mesh for Sample Problem

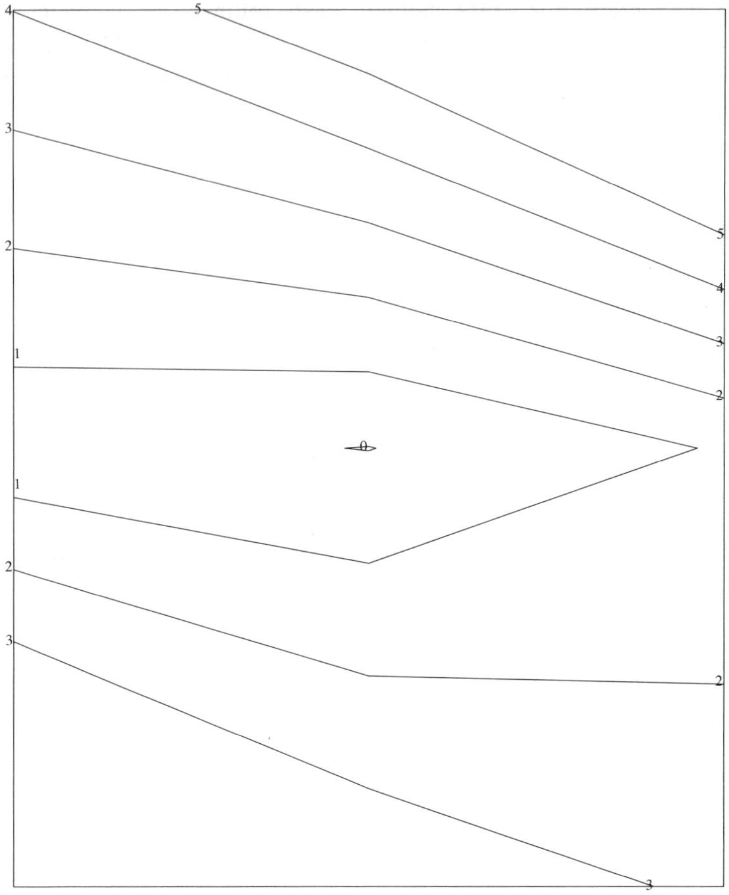

LEGEND FOR FRAME NUMBER 1 (STRESS: X) AT LOAD STEP 10

MINIMUM = -0.2197E+02 MAXIMUM = 0.1529E+03

0 = -0.2153E+02 1 = -0.2488E+01 2 = 0.1656E+02 3 = 0.3561E+02
4 = 0.5465E+02 5 = 0.7370E+02 6 = 0.9275E+02 7 = 0.1118E+03
8 = 0.1308E+03 9 = 0.1499E+03

Fig. 3.2. Stress σ_x Contours for Sample Problem

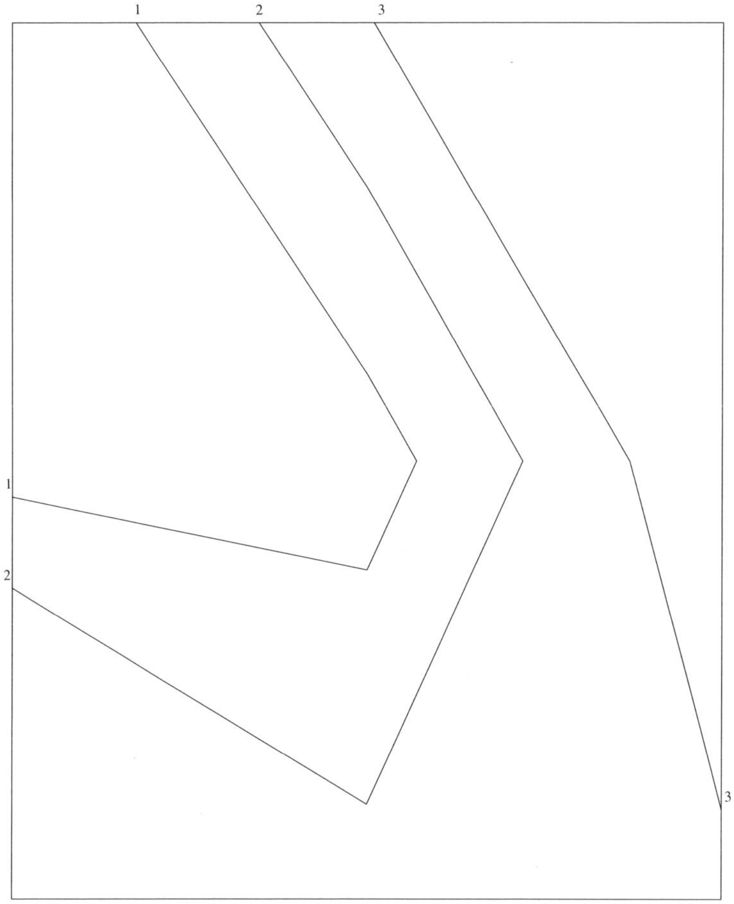

LEGEND FOR FRAME NUMBER 2 (STRESS: Y) AT LOAD STEP 10

MINIMUM = -0.1828E+02 MAXIMUM = 0.1557E+03

0 = -0.1791E+02 1 = 0.1036E+01 2 = 0.1998E+02 3 = 0.3893E+02
4 = 0.5787E+02 5 = 0.7682E+02 6 = 0.9577E+02 7 = 0.1147E+03
8 = 0.1337E+03 9 = 0.1526E+03

Fig. 3.3. Stress σ_y Contours for Sample Problem

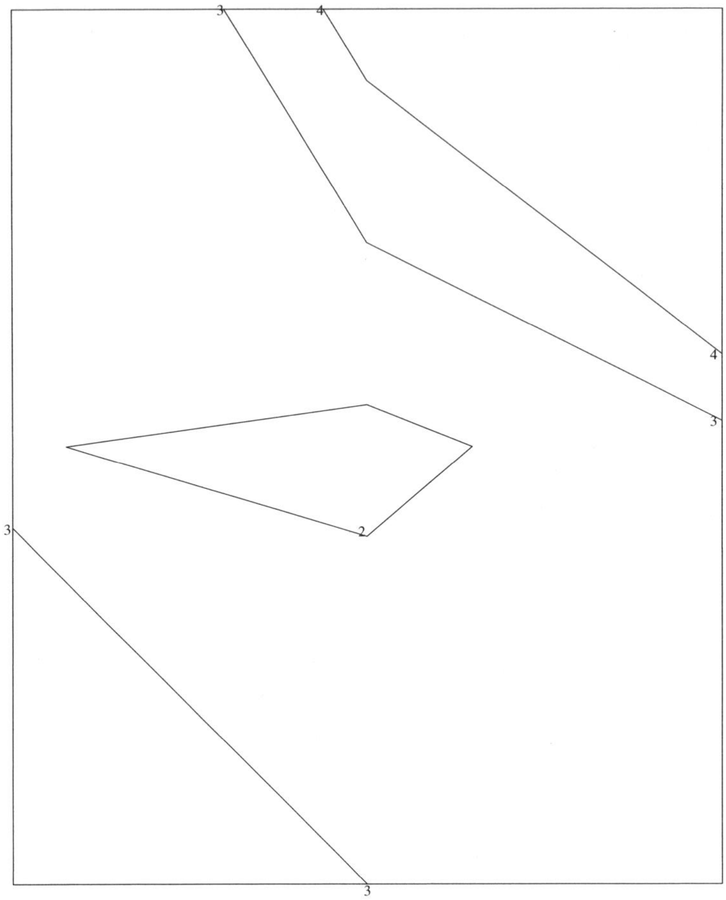

LEGEND FOR FRAME NUMBER 3 (STRESS: XY) AT LOAD STEP 10

MINIMUM = 0.5811E+01 MAXIMUM = 0.8349E+02

0 = 0.5928E+01 1 = 0.1436E+02 2 = 0.2279E+02 3 = 0.3123E+02
4 = 0.3966E+02 5 = 0.4809E+02 6 = 0.5653E+02 7 = 0.6496E+02
8 = 0.7339E+02 9 = 0.8182E+02

Fig. 3.4. Stress τ_{xy} Contours for Sample Problem

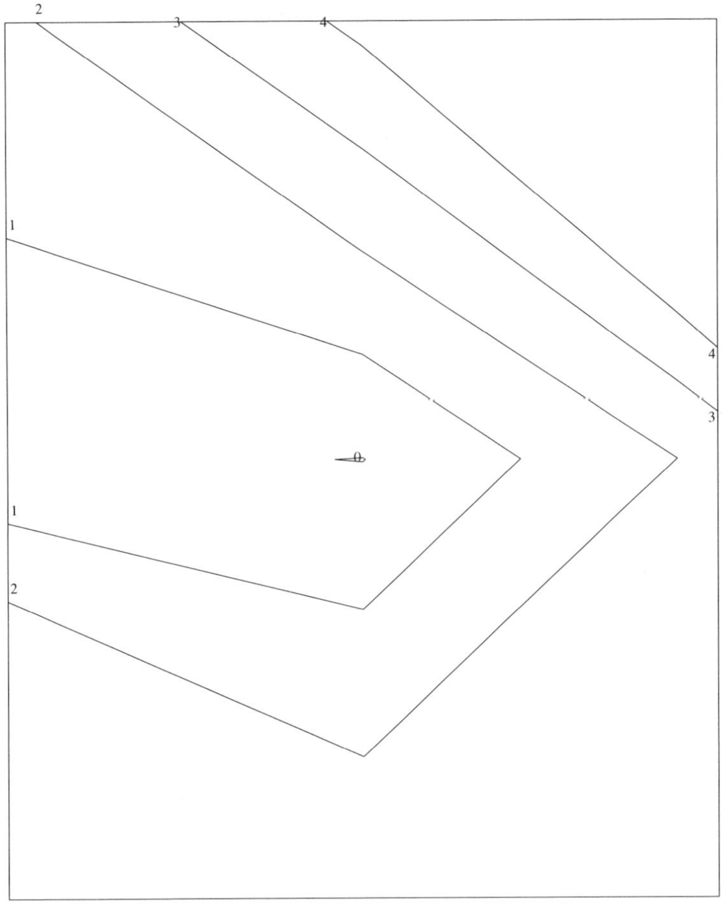

LEGEND FOR FRAME NUMBER 4 (STRESS: Z) AT LOAD STEP 10

MINIMUM = -0.8925E+01 MAXIMUM = 0.9260E+02

0 = -0.8746E+01 1 = 0.2309E+01 2 = 0.1336E+02 3 = 0.2442E+02
4 = 0.3547E+02 5 = 0.4653E+02 6 = 0.5758E+02 7 = 0.6864E+02
8 = 0.7969E+02 9 = 0.9075E+02

Fig. 3.5. Stress σ_z Contours for Sample Problem

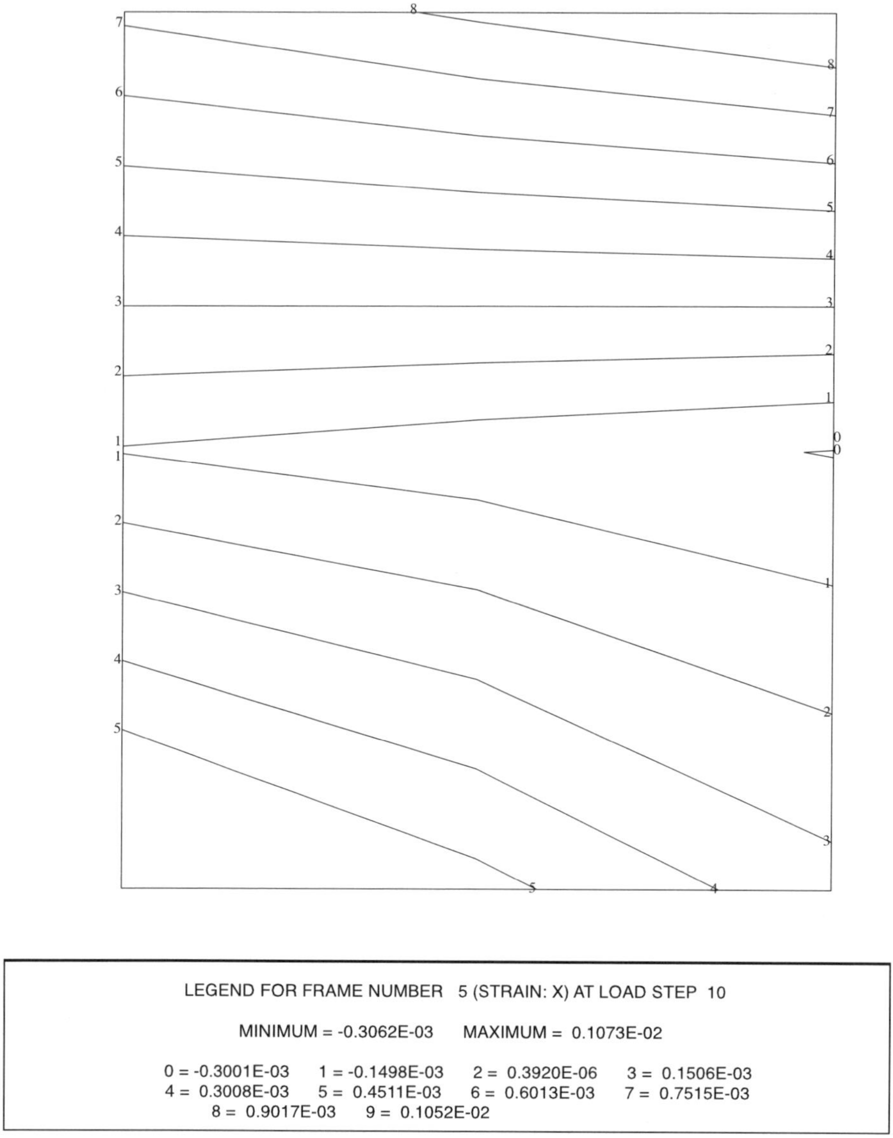

Fig. 3.6. Strain ε_x Contours for Sample Problem

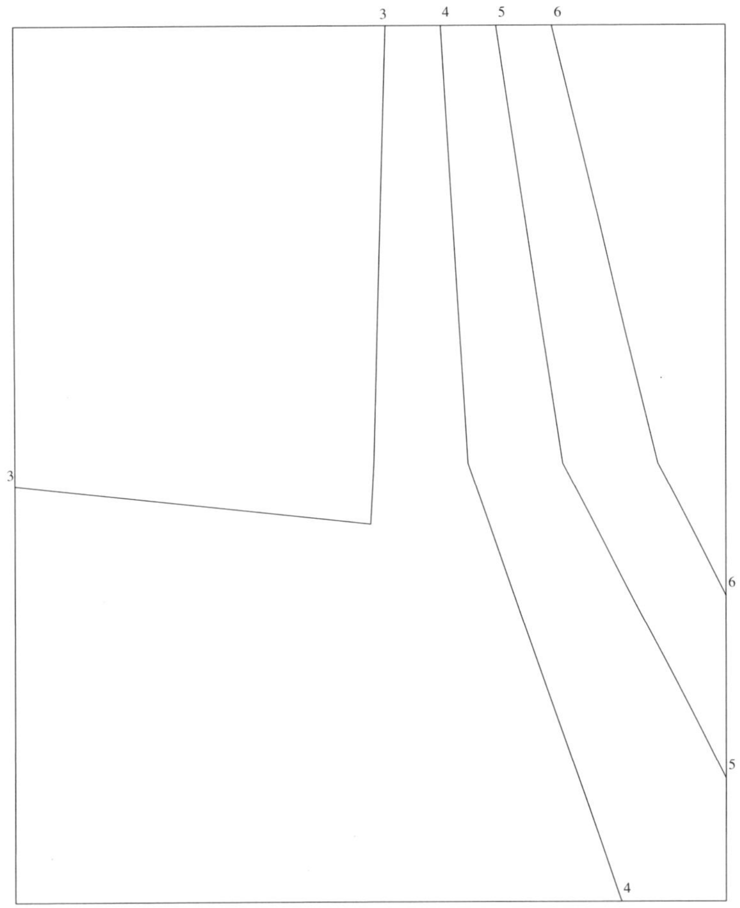

LEGEND FOR FRAME NUMBER 6 (STRAIN: Y) AT LOAD STEP 10

MINIMUM = -0.5205E-03 MAXIMUM = 0.1123E-02

0 = -0.5100E-03 1 = -0.3311E-03 2 = -0.1522E-03 3 = 0.2672E-04
4 = 0.2056E-03 5 = 0.3846E-03 6 = 0.5635E-03 7 = 0.7424E-03
8 = 0.9213E-03 9 = 0.1100E-02

Fig. 3.7. Strain σ_y Contours for Sample Problem

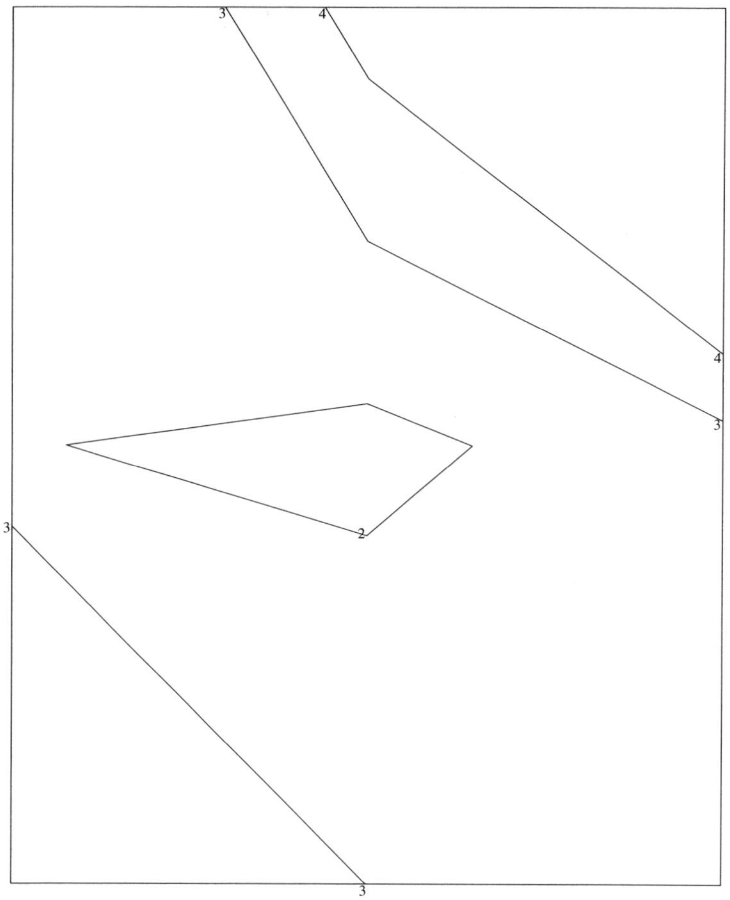

LEGEND FOR FRAME NUMBER 7 (STRAIN: XY) AT LOAD STEP 10

MINIMUM = 0.2067E-03 MAXIMUM = 0.2970E-02

0 = 0.2109E-03 1 = 0.5109E-03 2 = 0.8109E-03 3 = 0.1111E-02
4 = 0.1411E-02 5 = 0.1711E-02 6 = 0.2011E-02 7 = 0.2311E-02
8 = 0.2611E-02 9 = 0.2911E-02

Fig. 3.8. Strain γ_{xy} Contours for Sample Problem

LEGEND FOR FRAME NUMBER 8 (STRAIN: Z) AT LOAD STEP 10

MINIMUM = 0.0000E+00 MAXIMUM = 0.0000E+00

0 = 0.0000E+00 1 = 0.0000E+00 2 = 0.0000E+00 3 = 0.0000E+00
4 = 0.0000E+00 5 = 0.0000E+00 6 = 0.0000E+00 7 = 0.0000E+00
8 = 0.0000E+00 9 = 0.0000E+00

Fig. 3.9. Strain ε_z Contours for Sample Problem

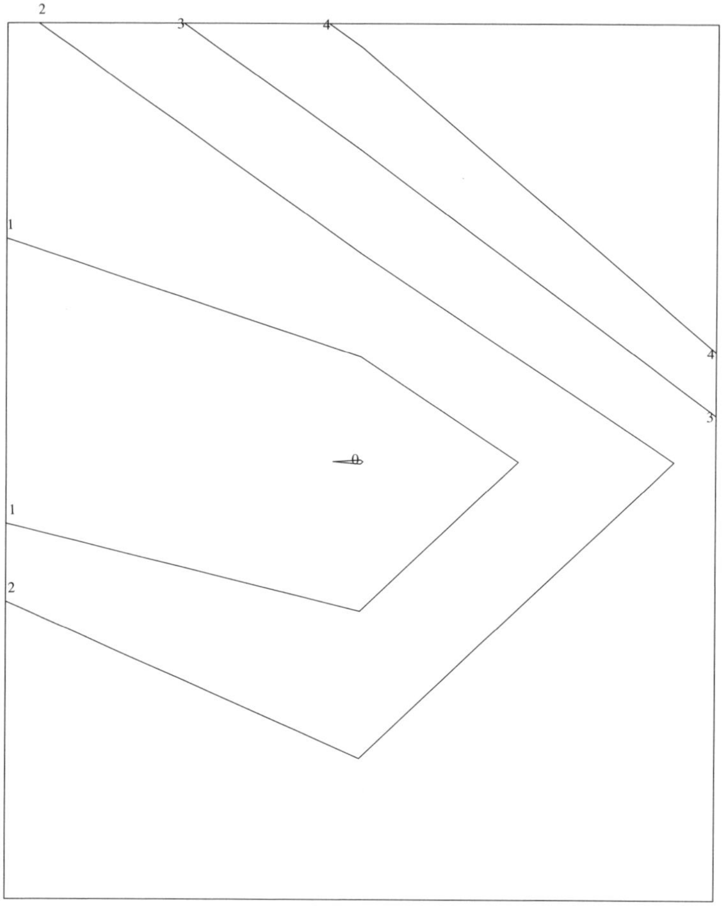

LEGEND FOR FRAME NUMBER 9 (VOLUMETRIC STRAINS) AT LOAD STEP 10

MÍNIMUM = -0.1289E+02 MAXIMUM = 0.1338E+03

0 = -0.1263E+02 1 = 0.3335E+01 2 = 0.1930E+02 3 = 0.3527E+02
4 = 0.5124E+02 5 = 0.6721E+02 6 = 0.8318E+02 7 = 0.9914E+02
8 = 0.1151E+03 9 = 0.1311E+03

Fig. 3.10. Volumetric Strain ε_v Contours for Sample Problem

LEGEND FOR FRAME NUMBER 10 (WORK) AT LOAD STEP 10

MINIMUM = 0.0000E+00 MAXIMUM = 0.0000E+00

0 = 0.0000E+00 1 = 0.0000E+00 2 = 0.0000E+00 3 = 0.0000E+00
4 = 0.0000E+00 5 = 0.0000E+00 6 = 0.0000E+00 7 = 0.0000E+00
8 = 0.0000E+00 9 = 0.0000E+00

Fig. 3.11. Plastic Work κ Contours

3.5 Installing the DNA Windows Interface

A Winodws program (or interface) is also available so that the user can edit the input file, run the analysis, view the output file, print the results, and view the finite element mesh and contours, all from within the same Windows environment. Here are the instructions to install the DNA Windows interface successfully on the hard disk:

1. Make a folder on the C: drive of your computer hard disk and call it DNA. This must be done on the C: partition of your hard disk and not on any other partition for the Windows interface to work properly. You should now have the folder C:\DNA available on your computer.
2. Copy the following files from the accompanying CD-ROM into the folder C:\DNA that you created in step 1.

   DNA.EXE

   DNA_S.EXE

   DNA_L.EXE

   SAMPLE.IN

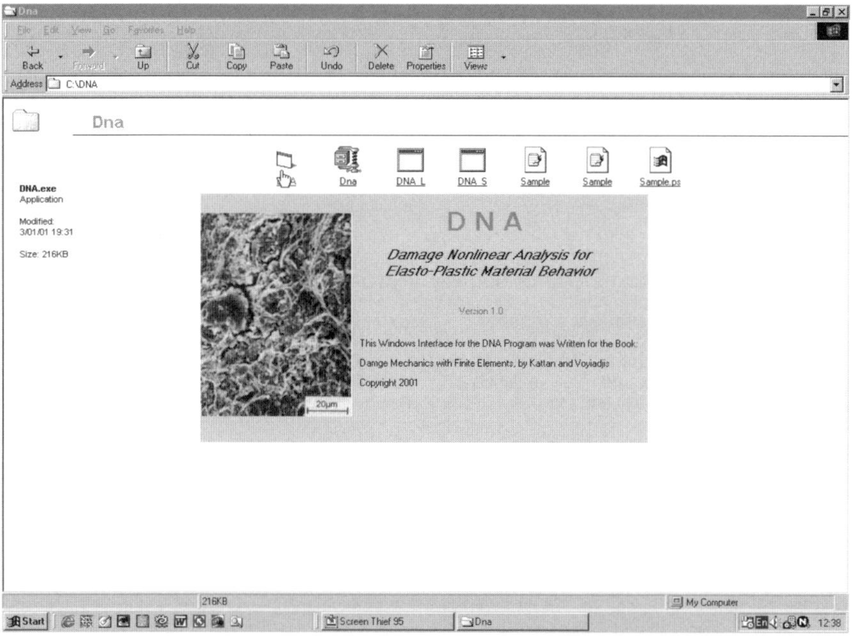

Fig. 3.12. Splash Screen of DNA Windows Interface

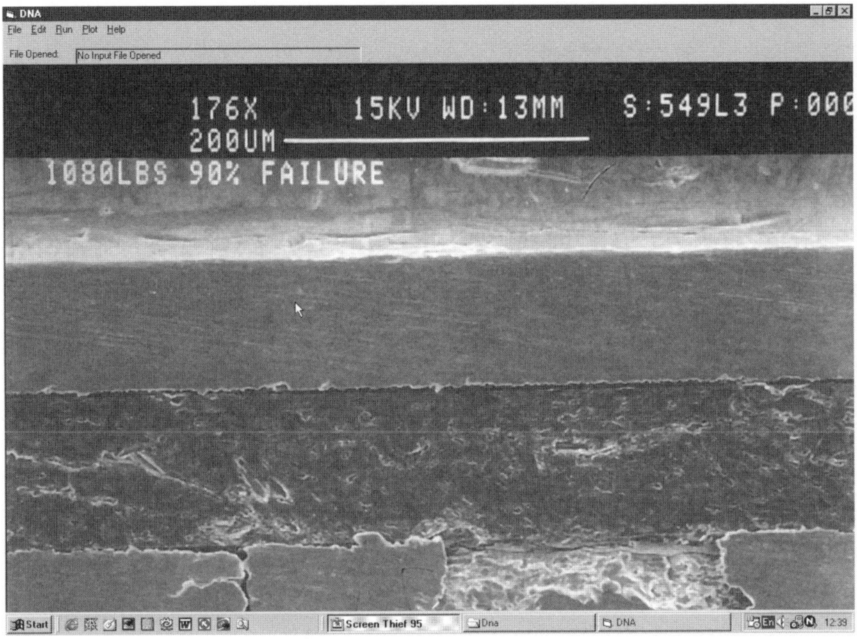

Fig. 3.13. DNA Windows Interface with Menu Bar

3. Make sure that you have the Ghostscript program GSView installed on your computer in a folder called C:\GSTOOLS. See Section 3.3 to install it if it is not already installed.
4. Run the program DNA.EXE from within Windows. This will start the DNA Windows interface. The splash screen will appear for a few seconds as shown in Figure 3.12. Then the interface will start with one menu bar as shown in Figure 3.13. See Section 3.6 below for instructions on how to use the interface.

3.6 Using the DNA Windows Interface

The DNA Windows interface is an integrated environment where you can perform all the operations of editing the input file, running the analysis, viewing the results and contours all from within the same menu bar. After installing the Windows DNA interface successfully, here are the instructions on how to use it:

1. Click the *File* menu and select *Open* in order to open an input file. You should get the name of the input file displayed at the top left corner of the screen.

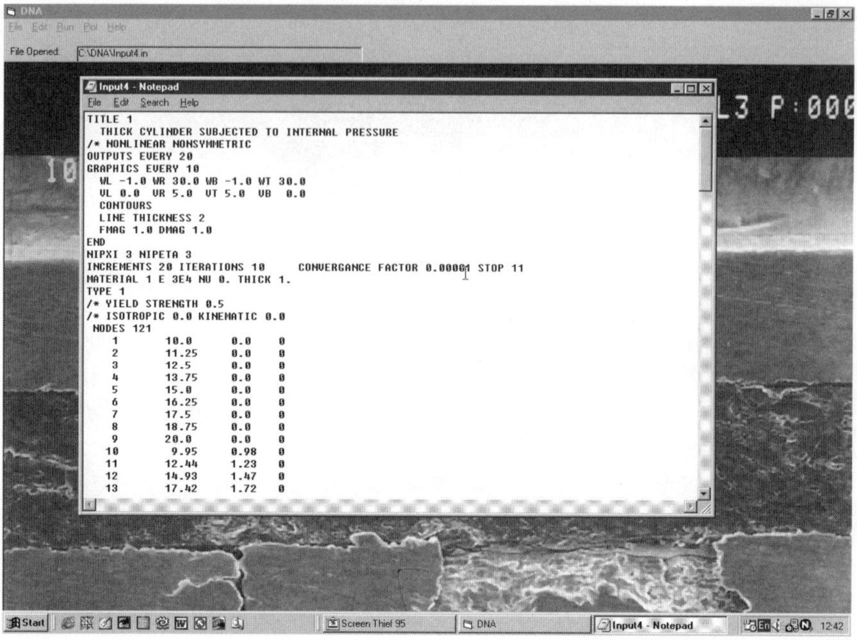

Fig. 3.14. Editing the Input File with NotePad

2. Click the *Edit* menu and select *Input File* to edit the input file. This will start the Winodws NotePad program and opens the desired input file as shown in Figure 3.14. If the input file is too long, the program will prompt you to start WordPad instead. To open a new file, open first an existing file then click the *File* menu and select *New* for the NotePad menu bar.

3. Click the *Run* menu and select the type of analysis you would like to run. Both small strain and large strain damage analysis are available at this step. Once the desired type is selected, Windows will start either DNA_S or DNA_L in a DOS Window (see Figure 3.15) depending on the type selected.

4. Click the *Edit* menu and select *Output File* in order to view the results in the output file. This will again start NotePad and open the output file. The output file is also available for editing at this step.

5. Click the *Plot* menu to view the finite element mesh and contours. This will start GSView automatically as shown in Figure 3.16. This step will run successfully only if the GRAPHICS option is included in the input file.

6. Click the *Help* menu for information on the program.

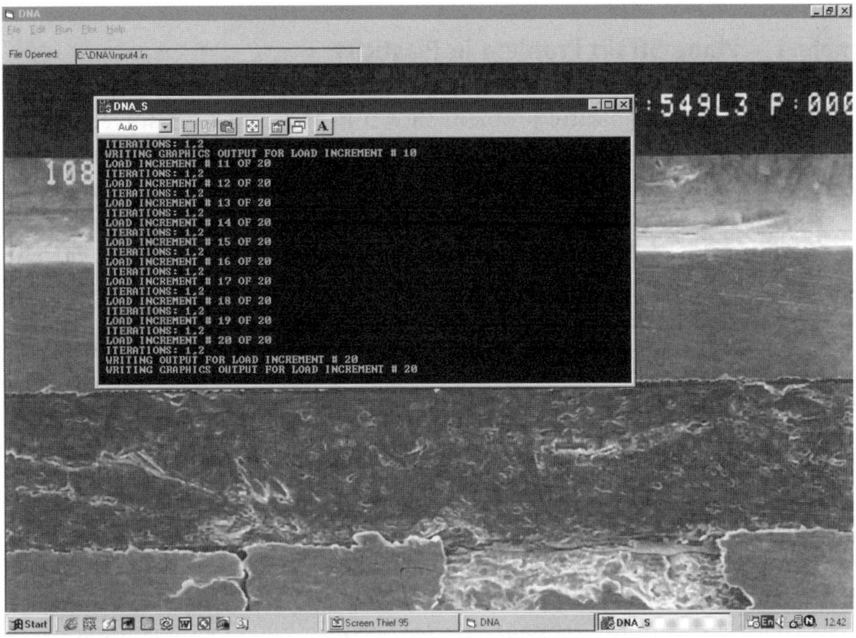

Fig. 3.15. Running the Analysis in a DOS Window

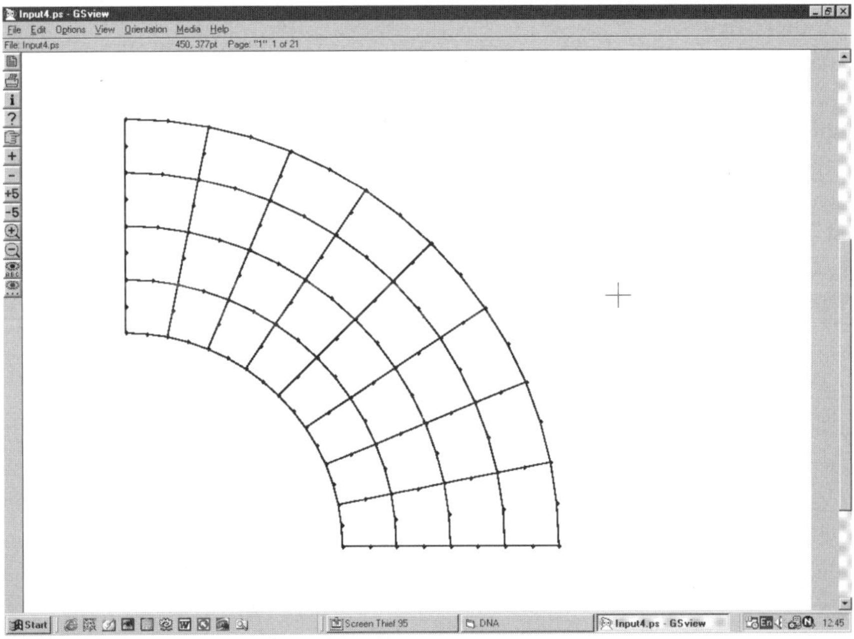

Fig. 3.16. Viewing the Finite Element Mesh with GSView

3.7 Example 1 – Plane Strain Problem in Plasticity

The plane strain elasto-plastic problem shown in Figure 3.17 is solved in this example using DNA. Two concentrated loads of 10 kN each act at the upper right corner (node 9) of the plate. It is assumed that plane strain conditions exist in the plate. The plate is made of aluminum alloy 2024-T3 with E = 73,089 *MPa* and $\nu = 0.3$. The thickness of the plate is 3.175 mm. The isotropic and kinematic hardening parameters used are c = 792.9 *MPa* and b = 275.8 *MPa*, respectively. Initial yielding is characterized by

$$\frac{\sigma_y}{\sqrt{3}} = 226.8\,MPa$$

Four-node isoparametric quadrilateral elements (element type 2204) are used to model this problem. Only four such elements are used for illustrative purposes. More elements should be used to obtain reliable results. A total of 100 increments of the load are used with 50 iterations per increment. A convergence factor of 0.0001 is set for this problem. This is obviously a TYPE 2 analysis (plasticity without damage effects) according to DNA terminology.

The DNA input file for this problem was shown previously in Table 3.2 for the sample example problem. In addition, graphical output including various stress and strain contours were shown in Figures 3.1–3.11 at the end of the first 10 increments.

3.8 Example 2 – Elastic Cantilever With Two Elements

This example is provided in order to check the accuracy of the results obtained by DNA and compare them with results of other finite element pro-

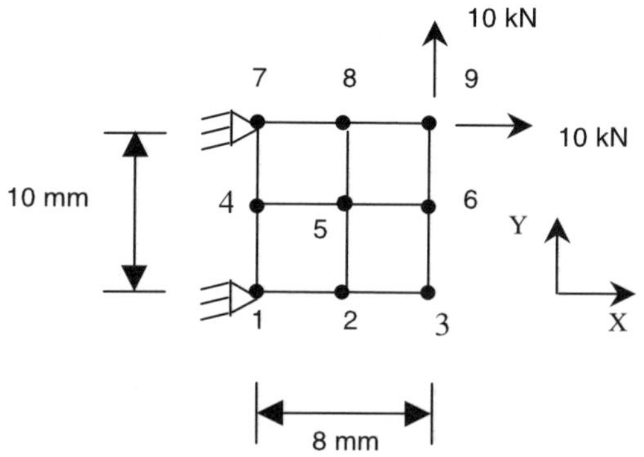

Fig. 3.17. Sample Problem for Example 1

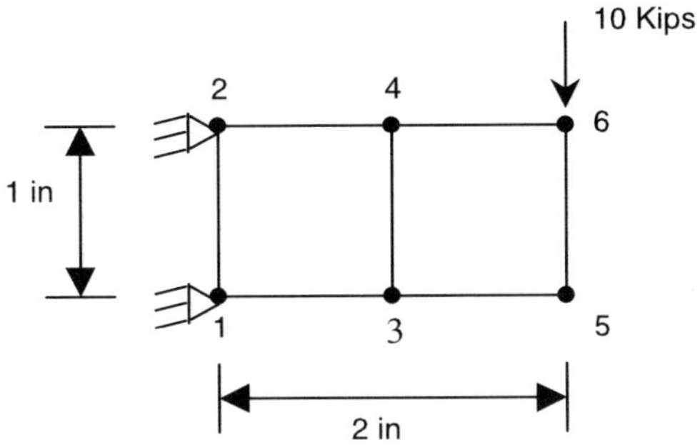

Fig. 3.18. Cantilever Beam with Two Elements

Table 3.3. Input File for Example 2 (test.in)

```
TITLE 1
TEST
NIPXI 3 NIPETA 3
MATERIAL 1 E 30000. NU 0.0 THICK 1.0
TYPE 1
NODES 6
1 0.0 0.0 0
2 0.0 1.0 0
3 1.0 0.0 0
4 1.0 1.0 0
5 2.0 0.0 0
6 2.0 1.0 0
BOUNDARY 6
1 1 1 0
2 1 1 0
3 0 0 0
6 0 0 1
INCIDENCES 2
1 2104,4, 1 3 4 2,1,0
2 2104,4, 3 5 6 4,1,0
LOADS 1
6 0.0 10.0 0.0
```

grams when no damage effects are present. A small elastic thin cantilever with two elements is considered as shown in Figure 3.18. Plane stress conditions are assumed to exist in this problem. The modulus of elasticity is $E = 30 \times 10^3$ ksi and Poisson's ratio is $\nu = 0.0$. The thickness of the cantilever plate is $t = 1$ in. Obviously this is a TYPE 1 analysis (elasticity with no damage effects) according to DNA terminology. The finite element mesh of two elements is shown in Figure 3.18. Only two elements are used for illustrative purposes. More elements should be used to obtain reliable results.

The cantilever plate is subjected to a concentrated load of 10 kips acting at its free edge (at node 6). Four-node isoparametric quadrilateral elements (element type 2104) are used. Three integration points are used in each direction for each element. The DNA input file for this problem is shown in Table 3.3.

After running this problem successfully with DNA, the output file shown in Table 3.4 is obtained. No graphical output is produced for this example as there are no graphics commands in the input file.

3.9 Example 3 – Center-Cracked Plate Under Uniaxial Tension

A center-cracked plate is analyzed as shown in Figure 3.19. The plate is subjected to uniaxial tension in the longitudinal direction. The material used is aluminum alloy 2024 T3 (E $= 73{,}087\,MPa$, $\nu = 0.3$) with both kinematic and isotropic hardening parameters of b $= 275.8\,MPa$ and c $= 792.9\,MPa$, respectively. Initial yielding is characterized by

$$\frac{\sigma_y}{\sqrt{3}} = 226.8\,MPa$$

Since the thickness of the plate (t $= 3.175\,mm$) is small compared with other dimensions, a state of plane stress is assumed.

Since the plate geometry and loading are symmetrical, only one quarter of the plate is discretized by finite elements (shaded area in the figure). Eight-node isoparametric quadrilateral finite elements are used in the finite element mesh as shown on page 428. We avoid the use of singularity elements around the crack tip by using a large number of regular elements at that point. The total number of elements used is 381 with 1228 nodes.

A load increment of $10\,MPa$ is used until a total load of $300\,MPa$ is reached. The DNA input file ME1PL.IN is available on the accompanying CD-ROM as well as the output file ME1PL.OUT. The finite element mesh (around the crack tip) for this problem is shown in Figure 3.20 which is taken from the file ME1PL.PS which can be converted into ME1PL.PDF using Adobe Acrobat Distiller. Both the postscript and PDF output files are available on the accompanying CD-ROM.

3.10 Compiling and Building the DNA Executables

In this Section, instructions are provided to compile and build the DNA executable files using Digital or Compaq Visual Fortran (version 5 or 6 or later) under Windows 95/98/NT or Windows 2000. The files can also be compiled using the old Microsoft Fortran Powerstation compiler. All the files needed are available on the accompanying CD-ROM. These instructions are provided

Table 3.4. Output File for Example 2 (test.out)

```
TEST

        COORDINATES OF THE NODES
 NODE NO.     X          Y          Z

  1  .0000000000   .0000000000   .0000000000
  2  .0000000000   1.000000000   .0000000000
  3  1.000000000   .0000000000   .0000000000
  4  1.000000000   1.000000000   .0000000000
  5  2.000000000   .0000000000   .0000000000
  6  2.000000000   1.000000000   .0000000000


       ELEMENT INCIDENCES (ELEMENT TYPE: 2104)
 ELEMENT NO.  INCIDENCES
     1   1 3 4 2
     2   3 5 6 4


         NODAL LOADS
 NODE NUMBER    X      Y     Z
   6     .0000     10.00    .0000

 NUMBER OF EQUATIONS =   8
 HALF BANDWIDTH =   8
 AVERAGE BANDWIDTH =   6
 SIZE OF THE STIFFNESS MATRIX =   36


 >>>>>>> OUTPUTS AT INCREMENT  1
        ********** ELEMENT= 1 **********
                     STRAIN COMPONENTS
 POINT  X      Y     EXX      EYY      EXY      EZZ
  1 .11270  .11270   1.54919E-03 -8.83935E-06  2.24624E-03  .00000
  2 .50000  .11270   1.54919E-03 -3.92157E-05  6.97043E-04  .00000
  3 .88730  .11270   1.54919E-03 -6.95920E-05 -8.52150E-04  .00000
  4 .11270  .50000   .00000  -8.83935E-06  2.21586E-03  .00000
  5 .50000  .50000   .00000  -3.92157E-05  6.66667E-04  .00000
  6 .88730  .50000   .00000  -6.95920E-05 -8.82527E-04  .00000
  7 .11270  .88730  -1.54919E-03 -8.83935E-06  2.18548E-03  .00000
  8 .50000  .88730  -1.54919E-03 -3.92157E-05  6.36290E-04  .00000
  9 .88730  .88730  -1.54919E-03 -6.95920E-05 -9.12903E-04  .00000
                          STRESS COMPONENTS
 POINT  X      Y     SXX      SYY      SXY      SZZ
  1 .11270  .11270   46.476   -.26518   33.694   .00000
  2 .50000  .11270   46.476   -1.1765   10.456   .00000
  3 .88730  .11270   46.476   -2.0878   -12.782   .00000
  4 .11270  .50000   .00000   -.26518   33.238   .00000
  5 .50000  .50000   .00000   -1.1765   10.000   .00000
```

Table 3.4. (continued)

```
 6 .88730  .50000  .00000  -2.0878  -13.238  .00000
 7 .11270  .88730  -46.476  -.26518  32.782  .00000
 8 .50000  .88730  -46.476  -1.1765  9.5444  .00000
 9 .88730  .88730  -46.476  -2.0878  -13.694  .00000
       ********** ELEMENT= 2 **********
                 STRAIN COMPONENTS
POINT  X     Y    EXX    EYY    EXY    EZZ
 1 1.1127  .11270  5.16398E-04 -1.65559E-05 9.70430E-04 .00000
 2 1.5000  .11270  5.16398E-04 1.96078E-04 4.54032E-04 .00000
 3 1.8873  .11270  5.16398E-04 4.08713E-04 -6.23655E-05 .00000
 4 1.1127  .50000  2.16840E-19 -1.65559E-05 1.18306E-03 .00000
 5 1.5000  .50000  2.16840E-19 1.96078E-04 6.66667E-04 .00000
 6 1.8873  .50000  2.16840E-19 4.08713E-04 1.50269E-04 .00000
 7 1.1127  .88730  -5.16398E-04 -1.65559E-05 1.39570E-03 .00000
 8 1.5000  .88730  -5.16398E-04 1.96078E-04 8.79301E-04 .00000
 9 1.8873  .88730  -5.16398E-04 4.08713E-04 3.62903E-04 .00000
                   STRESS COMPONENTS
POINT  X     Y    SXX    SYY    SXY    SZZ
 1 1.1127  .11270  15.492  -.49668  14.556  .00000
 2 1.5000  .11270  15.492  5.8824  6.8105  .00000
 3 1.8873  .11270  15.492  12.261  -.93548  .00000
 4 1.1127  .50000  6.50521E-15 -.49668  17.746  .00000
 5 1.5000  .50000  6.50521E-15 5.8824  10.000  .00000
 6 1.8873  .50000  6.50521E-15 12.261  2.2540  .00000
 7 1.1127  .88730  -15.492  -.49668  20.935  .00000
 8 1.5000  .88730  -15.492  5.8824  13.190  .00000
 9 1.8873  .88730  -15.492  12.261  5.4435  .00000
       REACTIONS AT THE NODES
NODE NO.   RX       RY      RZ
 1 -20.00000000  -4.705882353
 2  20.00000000  -5.294117647
 3 -2.1700306482E-16  -1.0172540436E-14
 4 -5.4928934665E-15  2.6562926121E-14
 5 -2.1250362581E-16  6.3901249743E-15
 6 7.5599249083E-15  10.00000000
       DISPLACEMENT OF THE NODES
NODE NO.   UX       UY      UZ
 1  .0000000000   .0000000000
 2  .0000000000   .0000000000
 3 2.0000000000E-03  2.7058823529E-03
 4 -2.0000000000E-03  2.6274509804E-03
 5 2.6666666667E-03  7.7647058824E-03
 6 -2.6666666667E-03  8.2352941176E-03

>>>>>>> TOTAL NUMBER OF ITERATIONS FOR THIS RUN IS =  2
CPU USAGE: 0:00:00.00 hrs
```

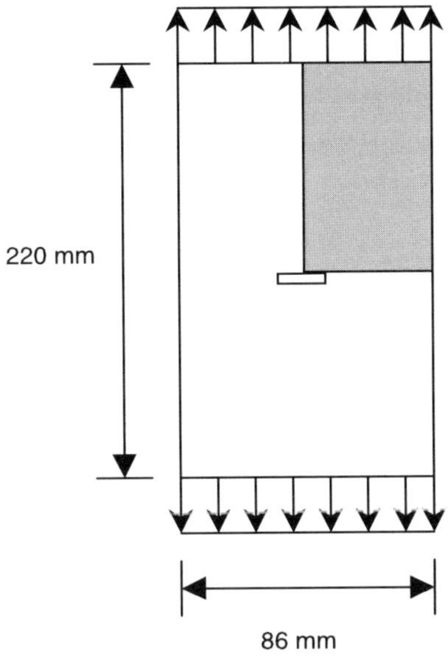

Fig. 3.19. Center-Cracked Plate Subjected to Uniaxial Tesnion

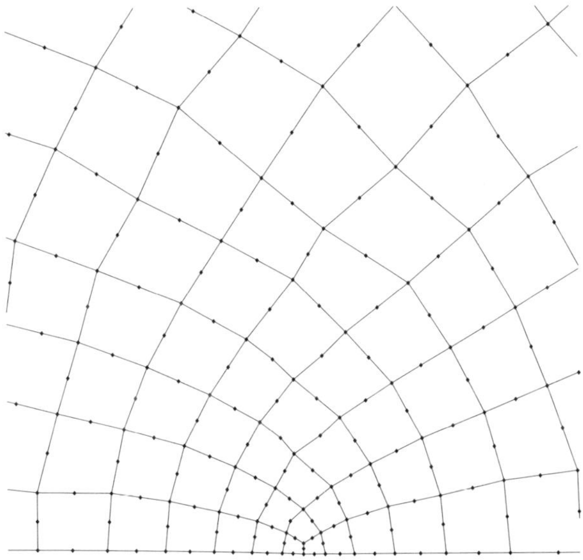

Fig. 3.20. Finite Element Mesh for Example 3

to programmers who wish to modify DNA or include additional features to the program.

In order to compile and build the file DNA_L.EXE, make a new project that includes the following files:

DNA_LARG.FOR
SXM.FOR
TENSOR.FOR
PRTL.FOR

The first file above is the source code for DNA_L while the last three files comprise the tensor library – a library of tensor operations that are extensively called by the program (see Chapter 5 for more details). You should now be able to build directly the executable file DNA_L.EXE.

In order to compile and build the file DNA_S, make a new project that includes the following files:

SOLUTION.FOR
GSTIFF.FOR
LOAD_VEC.FOR
ELEMLIB.FOR
BOUNDARY.FOR
CONTROL.FOR
TENS_UTL.FOR
PLOT.FOR
IO.FOR
UTILITY.FOR
STRS_VEC.FOR
DNA.FOR
BLOCKDAT.FOR
MATERIAL.FOR
SXM.FOR
TENSOR.FOR
PRTL.FOR

The first 14 files above contain the source code for DNA_S while the last three files comprise the tesnor library (see Chapter 5 for more details on the tensor library). You should now be able to build directly the executable file DNA_-S.EXE.

Finally, the programmer should note that the source code for DNA_L is contained in one large file while the source code for DNA_S is divided into 14 small files as shown above.

Exercises

1. Write a FORTRAN subroutine to evaluate the expression for damage evolution given in Equation (1.84). Use the tensor library.
2. Write a FORTRAN subroutine to evaluate the expression of Q given in exercise 1 of chapter 2. Use the tensor library.
3. Modify the FORTRAN source code of DNA (available on the accompanying CD-ROM) so as to include damage analysis of composite materials using the local approach. For details about the local approach see the book by Voyiadjis and Kattan (1999).
4. Modify the Visual Basic source code of the DNA Windows interface (available on the accompanying CD-ROM) so that the user can enter data interactively through the use of a graphical pre-processor instead of editing input files. The new Windows interface should include the capability of accepting data input interactively and writing it to an input file internally.

4 DNA Commands

DNA commands must be entered in the input file for any problem to be solved with this program. The following rules apply to DNA commands:

1. Real numbers can be entered using an "F" or "E" format with "D" or "E" exponent notation.
2. Many commands have numeric arguments but there are some commands that have no arguments.
3. Commands must be separated from their numeric arguments with white space (blank space or TAB character).
4. Numeric arguments can be separated by white space or comma.
5. Additionally, commands must be separated from previous commands (with or without arguments) with white space.
6. Anything following the characters /* on the line is considered a comment and will not be processed by DNA.
7. Only the first four characters of each command are read by DNA. The remaining part of the command is not processed.

4.1 Command Reference

The following is a comprehensive reference guide for all the known commands available in DNA. For each command, the format of the command, the types of arguments, notes, and examples are provided.

TITLE n

Followed by n lines for the title of the problem.

Notes: Usually n = 1, as one title line is sufficient. n is an integer.

Example:

TITLE 1
Analysis of a Cantilever Beam

LINEAR or NONLINEAR

Use LINEAR to perform a linear analysis and use NONLINEAR for a nonlinear analysis. Do not use both these commands in the same input file.

Example:

LINEAR

SYMMETRIC or NONSYMMETRIC

Use SYMMETRIC for a solution with a symmetric stiffness matrix. Use NONSYMMETRIC for a general nonsymmetric stiffness matrix. Do not use both these commands in the same input file.

Example:

NONSYMMETRIC

DIMENSION n

Use this command to set the dimension of the problem. Use n = 2 for two-dimensional problems and n = 3 for three-dimensional problems.

Note: n is an integer.

Example:

DIMENSION 2

TYPE n

Use this command to set the type of analysis required. Four analysis types are available in DNA as follows:

n = 1 for elastic analysis.
n = 2 for plastic analysis.
n = 3 for elastic damage analysis.
n = 4 for plastic damage analysis.

Note: n is an integer.

Example:

TYPE 2

COORDINATES n or NODES n or JOINTS n

Followed by n lines (or less when using node generation) of the following form:

k x_k y_k m	for two-dimensional problems
k x_k y_k z_k m	for three-dimensional problems.

where

n	number of nodes
k	node number
x_k	x-coordinate of node k
y_k	y-coordinate of node k
z_k	z-coordinate of node k (for three-dimensional problems only)
m	node number increment

Notes: When $m = 0$, all nodal coordinates must be given explicitly by the user. When $m \neq 0$, then all nodes between the start node and the end node will be generated using a node number increment of m.

The coordinates of the nodes must be entered using a consistent system of units.

The value of n should not exceed 3000.

n, k, m	are integers
x_k, y_k, z_k	are real numbers

Example: (for a two-dimensional problem)

NODES 6
1 0.0 0.0 0
2 0.0 1.0 0
3 1.0 0.0 0
4 1.0 1.0 0
5 2.0 0.0 0
6 2.0 1.0 0

MEMBERS n or INCIDENCES n

Followed by n lines (or less if element generation is used) of the form:

k l nl k_1 k_2 .... k_{nl} mat m

where

n	number of members or elements
k	element or member number

l	element type (see page 51 for element types)
nl	number of nodes per element (4, 8, or 20)
k_1 k_2 .... k_{nl}	node numbers of element k
mat	material type number of element k
m	element number increment

Notes: When $m = 0$, all element incidences must be given explicitly by the user. When $m \neq 0$, then all elements between the start element and the end element will be generated using an element number increment of m.

The value of n should not exceed 400, the value of mat should not exceed 10, and the value of nl should not exceed 20.

n, k, l, nl, k_1 k_2 .... k_{nl}, mat, m are integers

Example:

INCIDENCES 2

1 2104, 4, 1 3 4 2, 1, 0

2 2104, 4, 3 5 6 4, 1, 0

BOUNDARY n

Followed by n lines (or less if boundary generation is used) of the form:

k n_1 n_2 n_3 m

where

n	number of nodes where there are boundary conditions (e.g. supports)
k	node number where there is a boundary condition (e.g. support)
n_1 n_2 n_3	codes for boundary conditions (1 for a restrained degree of freedom, 0 for an unrestrained degree of freedom). Include n_3 only for three-dimensional problems.
m	increment number for boundary condition (e.g. support)

Note: n, k, n_1, n_2, n_3, m are integers

Example: (for a two-dimensional problem)

BOUNDARY 6

1 1 1 0

2 1 1 0

3 0 0 0

6 0 0 1

SKEW n

Followed by n lines of the form:
k l_1 l_2 l_3

where

n	number of nodes where there are skew (inclined) boundary conditions
k	node number where there is a skew (inclined) boundary condition
l_1 l_2 l_3	direction angles of skew (inclined) boundary conditions. Include l_3 only for three dimensional problems.

Notes: the value of n should not exceed 300.

n, k	are integers
l_1 l_2 l_3	are in degrees

Example:

SKEW 1
5 45 45 0

MATERIAL n E a NU b THICK c

Repeat this command line for each material type, where

n	material type number (should not exceed 10)
a	modulus of elasticity (Young's modulus)
b	Poisson's ratio
c	thickness of element (for two-dimensional problems only)

Notes: The values of a, b, and c must be entered using a consistent system of units

Example:

MATERIAL 1 E 73087.0 NU 0.3 THICK 3.175

LOADS n or LOADING n

Followed by n lines of the form:
k p_1 p_2 p_3

where

n	number of nodes where there are applied force loads
k	node number where there is an applied force load

p_1 p_2 p_3 applied force loads at each degree of freedom (should not exceed three applied force loads at each node). Include p_3 for three-dimensional problems only.

Notes: the values of p_1, p_2, and p_3 should be entered using a consistent system of units

n, k	are integers
p_1 p_2 p_3	are real numbers

Example: (for a three dimensional problem)

LOADS 1
6 0.0 10.0 0.0

DISPLACEMENTS n

Followed by n lines of the form:
k d_1 d_2 d_3

where

n	number of nodes where there are applied displacements
k	node number where there is an applied displacement
d_1 d_2 d_3	applied displacements at each degree of freedom (should not exceed three applied displacements at each node). Include d_3 for three-dimensional problems only.

Notes: The values of d_1, d_2, and d_3 should be entered using a consistent system of units

n, k	are integers
d_1 d_2 d_3	are real numbers

Example: (for a three dimensional problem)

DISPLACEMENTS 2
3 5.0 0.0 0.0
7 0.0 –7.4 0.0

NIPXI a NIPETA b NIPSI c

Use these commands to set the number of integration points for each element along the local axes, where:

a	number of integration points in the *xi*-direction
b	number of integration points in the η-direction
c	number of integration points in the ζ-direction. Include this value for three-dimensional problems only.

Notes: The values of a, b, and c should not exceed 3
a, b, c are integers

Example: (for a two-dimensional problem)

NIPXI 3 NIPETA 3

IRONS n

Use this command to set the optimum integration points introduced by Irons, where:

n number of Irons integration points

Note: n is an integer

Example:

IRONS 2

THICKNESS t

Use this command for entering the thickness of two-dimensional finite elements (also used as part of the MATERIAL command – see page 41), where:

t thickness of two-dimensional finite elements

Notes: The value of t must be entered using a consistent system of units
t is a real number

Example:

THICKNESS 3.175

E a

Use this command for entering the modulus of elasticity or Young's modulus of the material used (also used as part of the MATERIAL command), where:

a modulus of elasticity (Young's modulus)

Notes: The value of a must be entered using a consistent system of units
a is a real number

Example:

E 73087.0

NU b

Use this command for entering Poisson's ratio (also used as part of the MATERIAL command - see page 41), where:

b	Poisson's ratio

Notes: The value of b must be entered using a consistent system of units

b	is a real number

Example:

NU 0.3

WX a WY b WZ c

Use these commands for entering the specific weight of the material in each direction, where:

a	specific weight in x-direction
b	specific weight in y-direction
c	specific weight in z-direction (used only for three-dimensional problems)

Notes: The values of a, b, and c must be entered using a consistent system of units

a, b, c	are real numbers

Example:

WY 25.3

YIELD y

Use this command to set the yield strength of the material. Used only for plastic analysis (TYPE 2) and plastic damage analysis (TYPE 4), where:

y	yield strength of the material

Notes: The value of y must be entered using a consistent system of units

y	is a real number

Example:

YIELD 190.53

ISOTROPIC c

Use this command to set the isotropic hardening parameter of the material. Used only for plastic analysis (TYPE 2) and plastic damage analysis (TYPE 4), where:

c isotropic hardening parameter of the material

Notes: The value of c must be entered using a consistent system of units

c is a real number

Example:

ISOTROPIC 792.92

KINEMATIC a

Use this command to set the kinematic hardening parameter of the material. Used only for plastic analysis (TYPE 2) and plastic damage analysis (TYPE 4), where:

a kinematic hardening parameter of the material

Notes: The value of a must be entered using a consistent system of units

a is a real number

Example:

KINEMATIC 275.8

BETA b

Use this command to set the control parameter for kinematic hardening. Used only for plastic analysis (TYPE 2) and plastic damage analysis (TYPE 4), where:

b control parameter for kinematic hardening (1 for kinematic hardening, 0 for no kinematic hardening)

Note: b is an integer

Example:

BETA 1

DAMAGE d

Use this command to set the damage parameter of the material. Used only for elastic damage analysis (TYPE 3) and plastic damage analysis (TYPE 4), where:

d	damage parameter of the material

Notes: The value of d must be entered using a consistent system of units

d	is a real number

Example:

DAMAGE 0.4

COEFFICIENT c

Use this command to set the damage coefficient of the material. Used only for elastic damage analysis (TYPE 3) and plastic damage analysis (TYPE 4), where:

c	damage coefficient of the material

Notes: The value of c must be entered using a consistent system of units

c	is a real number

Example:

COEFFICIENT 2.0

DBBETA b

Use this command to set the partial derivative constant $\partial B/\partial \beta$ of the material. Used only for elastic damage analysis (TYPE 3) and plastic damage analysis (TYPE 4), where:

b	partial derivative constant $\partial B/\partial \beta$ of the material

Notes: The value of b must be entered using a consistent system of units

b	is a real number

Example:

DBBETA 18000000.0

INCREMENTS r

Use this command to set the number of increments for the analysis, where:

r	number of increments for the analysis

Note: r is an integer

Example:

INCREMENTS 50

ITERATIONS t

Use this command to set the number of iterations within each increment for the analysis, where:

t number of iterations within each increment

Note: t is an integer

Example:

ITERATIONS 20

CONVERGENCE c

Use this command to set the tolerance (acceptable error or convergence factor) for the analysis, where:

c tolerance (acceptable error or convergence factor) for the analysis

Note: c is a real number

Example:

CONVERGENCE 0.0001

FACLOW l

Use this command to set the lower convergence factor for the analysis, where:

l lower convergence factor for the analysis

Note: this command may be outdated in the latest version of DNA
l is a real number

Example:

FACLOW 0.000001

FACHIGH h

Use this command to set the higher convergence factor for the analysis, where:

h higher convergence factor for the analysis

Note: this command may be outdated in the latest version of DNA
h is a real number

Example:

FACHIGH 0.001

STOP n

Use this command to stop processing if divergence occurs after n iterations, where:

n number of iterations to stop processing if divergence occurs

Note: n is an integer

Example:

STOP 30

RESTART

Use this command to restart the analysis.

Example:

RESTART

OUTPUT EVERY n

Use this command to request writing the results to the output file every n increments, where:

n number of increments for output

Note: n is an integer

Example:

OUTPUT EVERY 10

GRAPHICS EVERY n

WL a WR b WB c WT d

FMAG p DMAG q

CONTOURS

LINE THICKNESS l

END

Use this command block to request plotting the finite element mesh and contours to a postscript file every n increments, where:

n	number of increments for plotting
a	lower left x-coordinate of graphics window
b	upper right x-coordinate of graphics window
c	lower left y-coordinate of graphics window
d	upper right y-coordinate of graphics window
p	magnification factor for original geometry
q	magnification factor for displacements
l	thickness of all lines to be drawn

Notes: Use CONTOURS only if you want contour plots

n, l	are integers
a, b, c, d, p, q	are real numbers

Example:

GRAPHICS EVERY 5
WL 0.0 WR 10.0 WB 0.0 WT 30.0
CONTOURS
LINE THICKNESS 1
FMAG 1.0 DMAG 1.0
END

There are a few other DNA commands that are undocumented and are not used in this manual. They are PAX, PAY, PBX, PBY, RAX, RAY, RBX, RBY, BLAN, and ENDF.

4.2
Library of Finite Elements in DNA

The following types of finite elements are supported by DNA. Each element description is given along with the element type number and a sketch showing the used node numbering scheme in DNA.

1. 4-noded isoparametric quadrilateral element: there are three types of this element as follows:
   2104 for two-dimensional plane stress problems.
   2204 for two-dimensional plane strain problems.
   2304 for two-dimensional axisymmetric problems.
2. 8-noded isoparametric quadrilateral element: there are three types of this element as follows:
   2108 for two-dimensional plane stress problems.
   2208 for two-dimensional plane strain problems.
   2308 for two-dimensional axisymmetric problems.
3. 9-noded Lagrangian element: there are three types of this element as follows:
   2109 for two-dimensional plane stress problems
   2209 for two-dimensional plane strain problems
   2309 for two-dimensional axisymmetric problems
4. 8-noded hexadedron solid element: there is only one type of this element as follows:
   3008 for three-dimensional problems
5. 20-noded hexahedron solid element: there is only one type of this element as follows:
   3020 for three-dimensional problems

Fig. 4.1. 4-noded isoparametric quadrilateral element

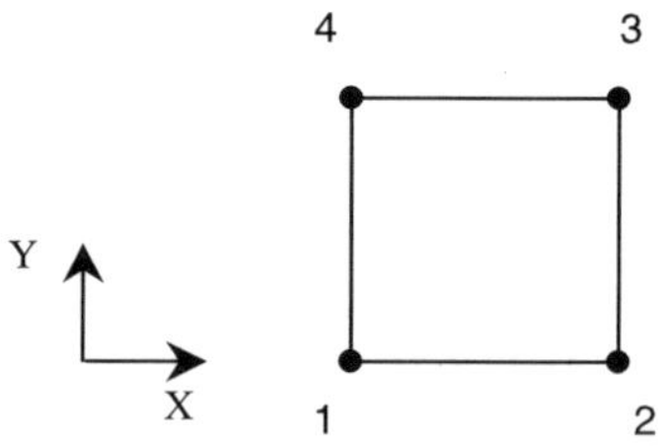

Fig. 4.2. 8-noded isoparametric quadrilateral element

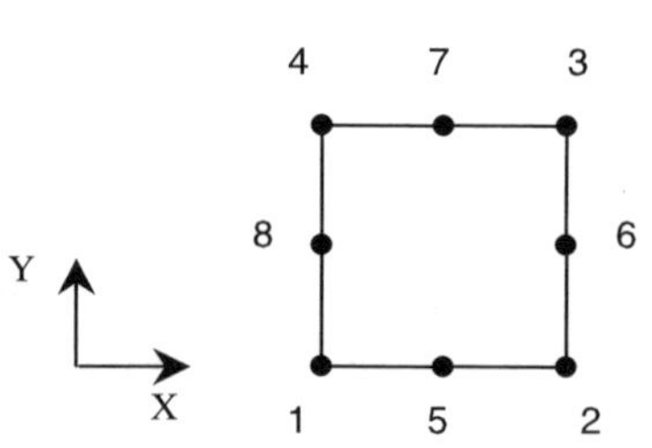

Fig. 4.3. 9-noded Lagrangian element

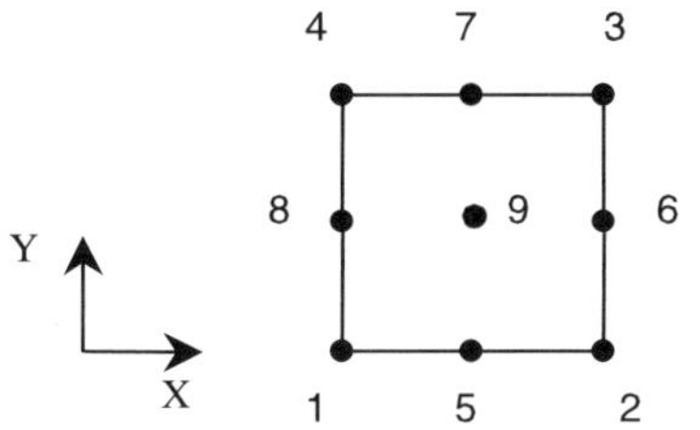

Fig. 4.4. 8-noded hexahedron solid element

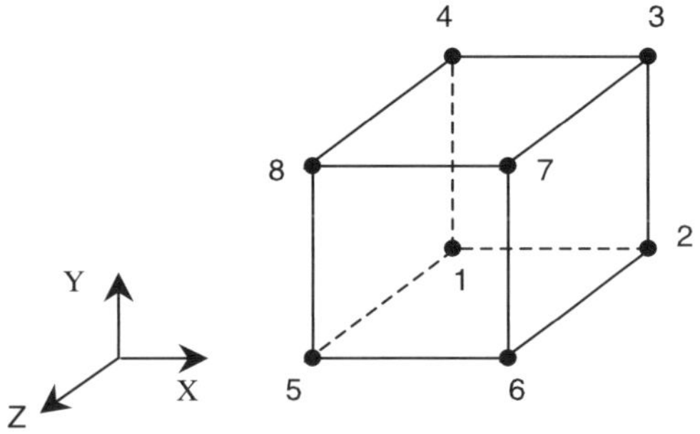

Fig. 4.5. 20-noded hexahedron solid element

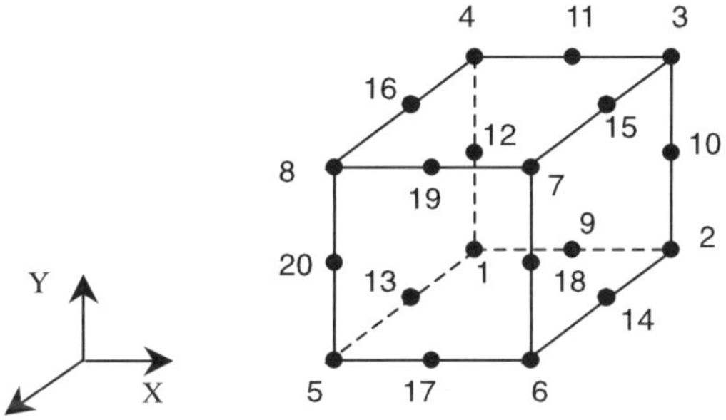

5 The Tensor Library

5.1 Introduction

The tensor library is a collection of 72 Fortran subroutines that handle several tensor operations useful in Damage Mechanics. The tensor library is called extensively by the DNA program to handle all the tensor operations. All tensors are assumed to have an index range of 1, 2, 3. All matrices are square of size 3×3 or 6×6 if full. The tensor library is available on the accompanying CD-ROM in three files, namely TENSOR.FOR, PRTL.FOR, and SXM.FOR. The subroutines are available for both single precision and double precision calculations. The following is a list of all the subroutines in the tensor library. Remove the letter "D" from the beginning of the name of each subroutine to use the single precision version; otherwise the double precision version will be used.

1. DAi_minus_Bi
2. DAi_plus_Bi
3. DAij_minus_Bij
4. DAij_plus_Bij
5. DAi_Bi
6. DAi_Bj
7. DAij_Bij
8. DAik_Bkj
9. DAij_Bkl
10. DAik_Bjl
11. DAil_Bjk
12. DAij_Bj
13. DAijkl_Bklmn
14. DAijkl_Bijkl
15. DAijkl_Bklij
16. DAijkl_Bijmn
17. DAijkl_Bmnkl
18. DAijkl_Bkl

19. DAijkl_Bmn
20. DAijkl_Bij
21. DAijklmn_Bmn
22. DAijklmn_Bkl
23. DAijklmn_Bij
24. Diarray
25. Dtensor_to_matrix
26. Dvector_to_tensor
27. Dtensor_to_vector
28. Dvector_6_to_tensor
29. Dtensor_to_vector_6
30. Dvector_9_to_tensor
31. Dtensor_to_vector_9
32. Dsymmetrize
33. Dtensor_to_matrix_full
34. Dmatrix_full_to_tensor
35. Darray_copy
36. Dtensor_inverse
37. DAijkl_minus_Bijkl
38. DAijkl_plus_Bijkl
39. DAijklmn_minus_Bijklmn
40. DAijklmn_plus_Bijklmn
41. Dscalar_multiply_Ai
42. Dscalar_multiply_Aij
43. Dscalar_multiply_Aijkl
44. Dscalar_multiply_Aijklmn
45. DAik_Bjk
46. DAki_Bkj
47. DAji_Bj
48. DAijklmn_Bijkl
49. DAijklmn_Bijmn
50. DAijklmn_Bklmn
51. DAijklmn_Bklij
52. DAijklmn_Bmnij
53. DAijklmn_Bmnkl
54. Didentity
55. Didentity_4
56. Dammij
57. Dsdeviator
58. Dsdeviator_4
59. Daijij
60. Daiijj

61. Daii
62. DAijklmn_Bijpq
63. DAijklmn_Bpqij
64. DAijklmn_Bklpq
65. DAijklmn_Bpqkl
66. DAijklmn_Bmnpq
67. DAijklmn_Bpqmn
68. Dpermute
69. Dmatrix_transpose
70. Dvector_full_to_tensor
71. Dtensor_to_vector_full
72. SXM

5.2 Fortran Source Code

The complete Fortran source code for the tensor library is stored on the accompanying CD-ROM. A few comment statements are given at the beginning of each subroutine explaining the function of the subroutine and the arguments used. See the DNA source code on the CD-ROM for examples on how to call these subroutines.

References

Allen D H, Harris C E (1987a) A Thermomechanical Constitutive Theory for Elastic Composites with Distributed Damage - I. Theoretical Formulation. International Journal of Solids and Structures, Vol. 23, pp. 1301-1318.

Allen D H, Harris C E, Groves S E (1987b) A Thermomechanical Constitutive Theory for Elastic Composites with Distributed Damage - II. Application to Matrix Cracking in Laminated Composites. International Journal of Solids and Structures, Vol. 23, pp. 1319-1338.

Allen D H, Harris C E, Groves S E, Norvel R G (1988) Characteristics of Stiffness Loss in Crossply Laminates with Curved Matrix Cracks. Journal of Composite Materials, Vol. 22, pp. 71-80.

Bahei-El-Din YA, Dvorak G J (1989) A Review of Plasticity Theory of Fibrous Composite Materials. In Metal Matrix Composites: Testing, Analysis and Failure Modes. Edited by W. S. Johnson, ASTM STP 1032, pp. 103-129.

Chaboche J L (1988) Continuum Damage Mechanics. Part I: General Concepts. Journal of Applied Mechanics, Vol. 55, pp. 59-64.

Chaboche J L (1988) Continuum Damage Mechanics. Part II: Damage Growth, Crack Initiation and Crack Growth. Journal of Applied Mechanics, Vol. 55, pp. 65-72.

Chaw C L, Wang J (1988) A Finite Element Analysis of Continuum Damage Mechanics for Ductile Fracture. Int. J. Fracture, Vol. 38, pp. 83-102.

Chaw C L, Lu T J (1989) On Evolution Laws of Anisotropic Damage. Eng. Fracture Mech., Vol. 34, pp. 679-701.

Christensen R M (1988) Tensor Transformations and Failure Criteria for the Analysis of Fiber Composite Materials. Journal of Composite Materials, Vol. 22, pp. 874-897.

Christensen R M (1990) Tensor Transformations and Failure Criteria for the Analysis of Fiber Composite Materials - Part II: Necessary and Sufficient Conditions for Laminate Failure. Journal of Composite Materials, Vol. 24, pp. 796-800.

Crisfield M A (1981) Finite Element Analysis of Combined Material and Geometric Nonlinearities. In Nonlinear Finite Element Analysis in Structural Mechanics. Wunderlin, W., et al., ed., Springer Verlag, New York, pp. 325-338.

Dvorak G J, Bahei-El-Din Y A (1979) Elastic-Plastic Behavior of Fibrous Composites. Journal of the Mechanics and Physics of Solids, Vol. 27, pp. 51-72.

Dvorak G J, Bahei-El-Din Y A (1982) Plasticity Analysis of Fibrous Composites. Journal of Applied Mechanics, Vol. 49, pp. 327-335.

Dvorak G J, Bahei-El-Din Y A (1987) A Bimodal Plasticity Theory of Fibrous Composite Materials. Acta Mechanica, Vol. 69, pp. 219-241.

Dvorak G J, Laws N (1987) Analysis of Progressive Matrix Cracking in Composite Laminates - II. First Ply Failure. Journal of Composite Materials, Vol. 21, pp.309-329.

Dvorak G J, Laws N, Hejazi M (1985) Analysis of Progressive Matrix Cracking in Composite Laminates - I. Thermoelastic Properties of a Ply with Cracks. Journal of Composite Materials, Vol. 19, pp.216-234.

Hill R (1965) A Self-Consistent Mechanics of Composite Material. Journal of the Mechanics and Physics of Solids, Vol. 13, pp. 213-222.

Hill R (1972) On Constitutive Macro-variables for Heterogeneous Solids at Finite Strain. Proceedings of the Royal Society - London, Vol. A326, pp. 131-147.

Hopkins H G, Wang A J (1954) Load-Carrying Capacities for Circular Plates of Perfectly-Plastic Material with Arbitrary Yield Conditions. J. Mech. Phys. Solids, Vol. 3, pp. 117-129.

Johnson W, Mellor P B (1973) Engineering Plasticity. Von Nostrand Reinhold, London.

Kachanov L M (1958) On the Creep Fracture Time. Izv Akad. Nauk USSR Otd Tekh., Vol. 8, pp. 26-31 (in Russian).

Kachanov L M (1986) Introduction to Continuum Damage Mechanics. Martinus Nijhoff Publishers, Dordreaht.

Kattan P I, Voyiadjis G Z (1990) A Coupled Theory of Damage Mechanics and Finite Strain Elasto-Plasticity - Part I: Damage and Elastic Deformations. International Journal of Engineering Science, Vol. 28, No. 5, pp. 421-435.

Kattan P I, Voyiadjis G Z (1993a) Micromechanical Analysis of Damage in Uniaxially Loaded Unidirectional Fiber-Reinforced Laminae. International Journal of Solids and Structures, Vol. 30, No. 1, pp. 16-36.

Kattan P I, Voyiadjis G Z (1993b) Overall Damage and Elastoplastic Deformation in Fibrous Metal Matrix Composites. International Journal of Plasticity, Vol. 9, pp. 931-949.

Kattan P I, Voyiadjis G Z (1996) Damage-Plasticity in a Uniaxially Loaded Composite Lamina: Overall Analysis. International Journal of Solids and Structures, Vol. 33, No. 4, pp. 555-576.

Krajcinovic D (1979) Distributed Damage Theory of Beams in Pure Bending. J. Appl. Mech., ASME, Vol. 46, pp. 592-596.

Krajcinovic D (1983) Constitutive Equation for Damaging Materials. Journal of Applied Mechanics, Vol. 50, pp. 335-360.

Krajcinovic D (1984) Continuum Damage Mechanics. Applied Mechanics Reviews, Vol. 37, No. 1, pp. 1-6.

Krajcinovic D (1989) Damage Mechanics. Mechanics of Materials, Vol. 8, pp. 117-197.

Krajcinovic D and Foneska G U (1981) The Continuum Damage Theory for Brittle Materials. Journal of Applied Mechanics, Vol. 48, pp. 809-824.

Laws N, Dvorak G J (1987) The Effect of Fiber Breaks and Aligned Penny-shaped Cracks on the Stiffness and Energy Release Rates in Unidirectional Composites. International Journal of Solids and Structures, Vol. 23, No. 9, pp. 1269-1283.

Lee J W, Allen D H, Harris C E (1989) Internal State Variable Approach for Predicting Stiffness Reduction in Fibrous Laminated Composites with Matrix Cracks. Journal of Composite Materials, Vol. 23, pp. 1273-1291.

Lee H, Peng K, Wang J (1985) An Anisotropic Damage Criterion for Deformation Instability and its Application to Forming Limit Analysis of Metal Plates. Engineering Fracture Mechanics, Vol. 21, pp. 1031-1054.

Lehmann Th (1991) Thermodynamical Foundation of Large Inelastic Deformations of Solid Bodies Including Damage. Int. J. Plasticity, Vol. 7, pp. 79-98.

Lemaitre J (1984) How to Use Damage Mechanics. Nuclear Engineering and Design, Vol. 80, pp. 233-245.

Lemaitre J (1985) A Continuous Damage Mechanics Model for Ductile Fracture. Journal of Engineering Materials and Technology, Vol. 107, pp. 83-89.

Lemaitre J (1986) Local Approach of Fracture. Engineering Fracture Mechanics, Vol. 25, Nos. 5/6, pp. 523-537.

Lemaitre J, Cordebois J P, Dufailly J (1979) Sur le Couplage Endommagement - Elasticite. Compte-rendu a lÁcademie des Sciences, Paris, B.391.

Lene F (1986) Damage Constitutive Relations for Composite Materials. Engineering Fracture Mechanics, Vol. 25, Nos. 5/6, pp. 713-728.

Li M, Wu S, Dang Z (1990) Determination of the Ductile Plastic Damage Variable for the Annealed 30CrMnSi. Int. J. Fracture, Vol. 46, pp. R51-R54.

Mori T, Tanaka K (1973) Average Stress in Matrix and Average Elastic Energy of Materials with Misfitting Inclusions. Acta Metallurgica, Vol. 21, pp. 571-574.

Murakami S (1988) Mechanical Modeling of Material Damage. Journal of Applied Mechanics, Vol. 55, pp. 280-286.

Shen W, Raio B, Lee H (1985) A Crack-Damage Mechanics Model for Composite Laminate. Engineering Fracture Mechanics, Vol. 21, No. 5, pp. 1019-1029.

Shi G, Voyiadjis G Z (1991) Efficient and Accurate Four-node Quadrilateral C° Plate Bending Element Based on Assumed Strain Fields. Int. J. Num. Meth. Eng., Vol. 32, pp. 1041-1055.

Shi G, Voyiadjis G Z (1992) A Simple Nonlayered Finite Element for the Elasto-plastic Analysis of Shear Flexible Plates. Int. J. Num. Meth. Eng., Vol. 33, pp. 85-99.

Shi G, Voyiadjis G Z (1993) A Computational Model for FE Ductile Plastic Damage Analysis of Plate Bending. Journal of Applied Mechanics, ASME, Vol. 60, pp. 749-758.

Shi G, Voyiadjis G Z (1997) A New Damage Strain Energy Release Rate for Plastic Damage Analysis. Mechanics Research Communications Journal, Vol. 24, No. 4, pp. 377-383.

Sidoroff F (1981) Description of Anisotropic Damage Application to Elasticity. In IUTAM Colloquium on Physical Nonlinearities in Structural Analysis, pp. 237-244, Springer-Verlag, Berlin.

Sumarac D, Krajcinovic D (1989) A Mesomechanical Model for Brittle Deformation Processes: Part II. J. Appl. Mech., ASME, Vol. 56, pp. 57-62.

Talreja R (1985) A Continuum Mechanics Characterization of Damage in Composite Materials. Proceedings of the Royal Society - London, Vol. A399, pp. 195-216.

Talreja R (1986) Stiffness Properties of Composite Laminates with Matrix Cracking and Interior Delamination. Engineering Fracture Mechanics, Vol. 25, Nos. 5/6, pp. 751-764.

Tang L, Chen W, Liu Y (1980) Quasi-Conforming Elements for Finite Element Analysis. J. Dalian Inst. Tech. (in Chinese), Vol. 19, No. 2, pp. 19-36.

Voyiadjis G Z (1988) Degradation of Elastic Modulus in Elastoplastic Coupling with Finite Strains. Int. J. Plasticity, Vol. 4, pp. 335-353.

Voyiadjis G Z, Kattan P I (1990) A Coupled Theory of Damage Mechanics and Finite-Strain Elasto-Plasticity, Part II: Damage and Finite Strain Plasticity. International Journal of Engineering Science, Vol, 28, No. 6, pp. 505-524.

Voyiadjis G Z, Kattan P I (1992) A Plasticity-Damage Theory for Large Deformation of Solids - Part I: Theoretical Formulation. International Journal of Engineering Science, Vol. 30, No. 9, pp. 1089-1108.

Voyiadjis G Z, Kattan P I (1993) Damage of Fiber-Reinforced Composite Materials with Micromechanical Characterization. International Journal of Solids and Structures, Vol, 30, No. 20, pp. 2757-27778.

Voyiadjis G Z, Kattan P I (1999) Advances in Damage Mechanics: Metals and Metal Matrix Composites. Elsevier, Amsterdam.

Voyiadjis G Z, Park T (1997a) Anisotropic Damage Effect Tensor for the Symmetrization of the Effective Stress Tensor. Journal of Applied Mechanics, ASME, Vol 64, pp. 106-110.

Voyiadjis G Z, Park T (1997b) Local and Interfacial Damage Analysis of Metal Matrix Composites Using the Finite Element Method. Journal of Engineering Fracture Mechanics, Vol. 56, No. 4, pp. 483-511.

Voyiadjis G Z, Shi G (1991a) A Refined Two-Dimensional Theory for Thick Cylindrical Shells. Int. J. Solids Struct., Vol. 27, pp. 261-282.

Voyiadjis G Z, Shi G (1991b) Buckling and Damage Analysis of Plates and Shells. Report to Martin Marietta Manned Space Systems at New Orleans.

Voyiadjis G Z, Thiagarajan G (1996) A Cyclic Anisotropic Plasticity Model for Metal Matrix Composites. International Journal of Solids and Structures, Vol. 33, No. 4, pp. 555-576.

Voyiadjis G Z, Venson A, Kattan P I (1995) Experimental Determination of Damage Parameters in Uniaxially-Loaded Metal Matrix Composites Using the Overall Approach. International Journal of Plasticity, Vol. 11, No. 8, pp. 895-926.

Contents of the Accompanying CD-ROM

The accompanying CD-ROM includes five folders as follows:

Executables:

This folder includes the DNA DOS executables and the DNA Windows Interface. See Sections 3.1 and 3.2 for installing and running DNA.

Example Input Files:

This folder includes some example input files. Also the resulting output, postscript, and PDF files for some of these examples are included.

Ghostscript:

This folder includes the compressed file GSV27550.EXE which contains both Ghostscript and GSView. See Section 3.3 for installing and running GSView.

DNA Source Code:

This folder includes the complete source code for DNA in 18 Fortran files. These include the source code for DNA_S, DNA_L, and the tensor library. See Section 3.10 for compiling the source code and building the DNA executables.

DNA Windows Interface Source Code:

This folder includes the Visual Basic source code for the DNA Windows Interface. See Sections 3.5 and 3.6 for installing and running the DNA Windows Interface.

Subject Index

Printing (Computer to Film): Saladruck Berlin
Binding: Stürtz AG, Würzburg